This book is due for return not later than the
last date stamped below, unless recalled sooner.

Monographs in Electrical and Electronic Engineering

Editors: P. Hammond   T. J. E. Miller   S. Yamamura

Monographs in Electrical and Electronic Engineering

Spiral Vector Theory of AC Circuits and Machines

Sakae Yamamura

Member of Japan Academy
Professor Emeritus, University of Tokyo
Honorary Member of IEE (Japan), Life Fellow of IEEE (USA)
Dr. Engr., Ph.D., M.S., B.E.

CLARENDON PRESS · OXFORD

1992

Oxford University Press, Walton Street, Oxford OX2 6DP

Oxford New York Toronto
Delhi Bombay Calcutta Madras Karachi
Petaling Jaya Singapore Hong Kong Tokyo
Nairobi Dar es Salaam Cape Town
Melbourne Auckland

and associated companies in
Berlin Ibadan

Oxford is a trade mark of Oxford University Press

Original Japanese edition published as 'Koryu Mota no Kaiseki to
Seigyo' by Sakae Yamamura
© Sakae Yamamura, 1988, published by Ohmsha, Ltd., Tokyo, Japan

English edition © Oxford University Press, 1992
English translation rights arranged with Ohmsha, Ltd.

Published in the United States
by Oxford University Press, New York

A catalogue record for this book is available from the British Library

Library of Congress Cataloging in Publication Data
Yamamura, Sakae, 1918–
[Kōryu mōta no kaiseki to seigyo. English]
Spiral vector theory of AC circuits and machines / Sakae Yamamura.
p. cm. — (Monographs in electrical and electronic
engineering; 26)
Translation of: Kōryu mōta no kaiseki to seigyo.
Includes bibliographical references and index.
1. Induction motors. 2. Electric circuits—Alternating current.
3. Electric power system stability. I. Title. II. Series.
TK2785.Y3413 1991 621.319'13—dc20 91-37366
ISBN 0 19 859379 1

Typeset by Integral Typesetting, Gorleston, Norfolk
Printed and bound in Great Britain by
Biddles Ltd, Guildford and King's Lynn

Preface

The author taught electrical engineering and control theory for many years at the University of Tokyo, and tried to modernize electric machine engineering by combining it with automatic control theory. Automatic control engineering developed rapidly after the Second World War, during which its growth began. Control of electric machines has been modernized in combination with automatic control technology, and they are playing an ever increasing role in many areas of industry and transportation.

When utilization of electrical energy began at the end of the last century, it was generated and distributed as direct current. Alternating current appeared later, and for many years DC and AC were in serious competition. Today, the AC power system is more widely used. Nevertheless, DC motors are still used widely because they have better control performance, which makes them indispensable as control motors.

Progress in power electronics, however, has changed this situation. Advanced inverter techniques have made it possible to control not only the voltage and current but also the frequency of the AC power supply, and now AC quantities are controllable as AC vector quantities of good waveforms. AC motors were previously tied to a fixed synchronous speed, but they are now free from this tie, and their inherent superior performance can fully be utilized as control motors. As a consequence, AC motors are moving into application areas where DC motors were predominantly used. What constrained the controllability of AC motors was the poor controllability of the AC power supply.

Although freed from the constraint associated with the AC power supply, another obstacle still limited the development of AC motor applications under the new situation. This was the inadequate theory of AC motors. The author experienced this lack of adequate analytical theory of AC machines for many years. Transient phenomena associated with AC motors had not been analysed sufficiently, and adequate performance equations for them were not available. Because of this vacant area in AC motor analysis, it has not been possible to put AC motor control on a sound theoretical basis.

As an analytical theory of the AC motors, the perpendicular two-axis theory has been very popular, but it has not provided us with an adequate analytical solution. This theory is sometimes mistakenly called the d–q axis theory, and is confused with the d–q axis or two-reaction theory of the saliency of synchronous machines, causing problems in analysis and teaching.

The instantaneous value symmetrical component method[16] was proposed a little after Fortescue's symmetrical component method.[17] The latter method was very successful in analysing steady states in the asymmetrical, or unbalanced, operation of three-phase circuits and machines, and was and is still widely used. The former method was intended to analyse transients in the asymmetrical operation of three-phase systems. But it was not successful and has hardly been used. Its positive- and negative-sequence components are not independent and analysis in terms of these components is mathematically difficult and awkward, in spite of the variable transformation involved. The space vector, or space phasor, method was proposed, but this is nothing but the positive-sequence component of the instantaneous value symmetrical component method.[18, 19] In order to supplement the mathematical weak points, it resorts to physical pictures, which very much lack essential mathematical logic and proof. The physical pictures include superposition of the space plane, which is isotropic, and the complex plane, which is anisotropic. Their merger is mathematically and physically impossible. The vector method has rarely been successful in analysing transient phenomena in three-phase machines.[19] The vector control of AC motors,[25] which is essentially the same as field orientation or transvector control,[15] and was based on the two-axis method, is now switching to the space vector method, and has been and still is lacking a necessary analytical basis. This lack of adequate analytical theory has been an obstacle especially to fast control of AC motors used as control motors.

In order to overcome this obstacle, the author has proposed the spiral vector method of analysis. The spiral vector is an exponential function of time with a complex index: it can express both steady-state AC variables and transient-state variables, and even DC. When the state variables of three-phase machines are expressed as spiral vectors, the performance equation of the three-phase machine can be written in terms of variables of a single phase that represents the three phases on the stator or rotor. The author calls this method phase aggregation, and it has previously been used only in the steady-state analysis of three-phase machines where variables are expressed as phasors. The spiral vector now extends the phase aggregation method to transient analysis of three-phase machines, using the original variables without any variable transformation. The derived performance equations are easy to solve analytically, and their analytical solution reveals a very good understanding of electromagnetic transient phenomena. The control features thus extracted from three-phase motors are very much superior to those of DC motors.

The spiral vector method unifies steady-state AC circuit theory and transient-state circuit theory. The former uses phasor notation, while the latter uses instantaneous real values. The different variable expressions separate the two theories and make it awkward to combine their solutions. These two theories are now unified to establish spiral vector circuit theory. The spiral vector method extends applicability of the symmetrical component method to transient analysis of three-phase machines under asymmetrical operation. It seems that the spiral vector method will bring about revolutionary changes in AC circuit analysis.

The author has worked on spiral vector theory for about ten years and has published a number of original papers [1-13] and a book.[10] The theory still continues to grow and to find new applications. Recently, a new book on the subject by the author was published in Japanese.[20] It was well received in Japan. This is the English version.

Tokyo   S.Y.
June 1991

Contents

PART I
AC circuit theory

1 Spiral vector theory of AC circuits

It seems that alternating current theory needs revising in two respects. First, steady-state analysis and transient-state analysis are treated in two separate theories: AC circuit theory and transient theory. Second, transients of AC machines are not adequately and sufficiently analysed. In AC theory, vector, or phasor, notation has been used for steady-state analysis since the beginning of the century. Although phasor notation has some outstanding merits, its use is confined to steady-state analyses of AC circuits and machines. The phasor, or vector, is a stationary vector in the complex plane, which cannot express variables in a transient state. Transient-state variables are expressed by their instantaneous real values. Thus the steady-state theory of AC circuits is separated from the transient theory by different expressions for the state variables. It is cumbersome and awkward to combine steady and transient solutions obtained in two separate theories and expressed in different notations. The spiral vector method proposed in this book will unify the two theories by using a common expression: the spiral vector.

Transient phenomena of AC machines has remained an area lacking adequate and sufficient analysis. Progress of inverter techniques has made control of the AC power supply very quick and economical, and now AC motors are replacing DC motors in the wide area of motor drives. Transient phenomena in AC motors have thus gained more importance in motor operations, but their analysis has been very much neglected. The spiral vector method proposed in this book will promote AC motor transient analysis by providing adequate performance equations, as well as their analytical solutions.

Thus the spiral vector method will solve the two problems which have plagued AC circuit and machine theories throughout this century. The method will reform AC circuit and machine theories, to establish new analytical theories well-suited to modern applications of AC circuits and machines, where transient phenomena have greater importance.

1.1 Spiral vectors

Performance equations for electric circuits and machines can be written in the form of differential equations. Their general solutions can be written in terms

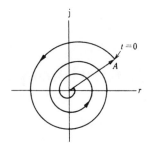

Fig. 1.1 A spiral vector in the complex plane

of exponential functions of time as follows:

$$i = A\, e^{\delta t}, \qquad \delta = -\lambda + j\omega. \tag{1.1}$$

When δ is complex, as in equation (1.1), as time progresses i depicts a spiral in the complex plane, as shown in Fig. 1.1. Let us call i a spiral vector.

As an example of the spiral vector method, we will use it to solve the following second-order differential equation:

$$a\frac{d^2 i}{dt^2} + b\frac{di}{dt} + ci = v. \tag{1.2}$$

Let the voltage v be expressed by

$$v = \sqrt{2}|V|\, e^{j(\omega t + \phi)} = \sqrt{2}V\, e^{j\omega t} = \sqrt{2}\dot{V}. \tag{1.3}$$

This is the spiral vector with $\lambda = 0$, and it depicts a circle in the complex plane. This spiral vector, therefore, will hereafter be called a circular vector. Here $\dot{V} = v/\sqrt{2} = V\, e^{j\omega t}$ is a circular vector and $V = |V|\, e^{j\phi}$ is a phasor. Let the solution of equation (1.2) be assumed to be a spiral vector as follows:

$$i = A\, e^{\delta t}. \tag{1.4}$$

For the steady state, δ becomes $j\omega$ and i becomes a circular vector. Then $di/dt = j\omega i$ and from equation (1.2) we get

$$[(j\omega)^2 a + (j\omega)b + c]i = \sqrt{2}|V|\, e^{j(\omega t + \phi)},$$

which gives the following steady-state current:

$$i_s = \frac{\sqrt{2}|V|}{-\omega^2 a + c + j\omega b}\, e^{j(\omega t + \phi)} = \sqrt{2}I\, e^{j\omega t} = \sqrt{2}\dot{I} \tag{1.5}$$

For the transient term, by inserting i of equation (1.4) into the left-side of equation (1.2), which we set equal to zero, we get

$$a\delta^2 + b\delta + c = 0. \tag{1.6}$$

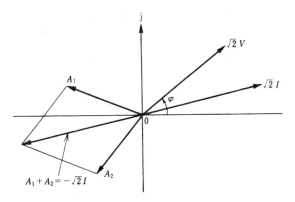

Fig. 1.2 Spiral vectors of equation (1.8) for $t = 0$

This is the characteristic equation, which gives two characteristic roots denoted δ_1 and δ_2. The general transient solution is then given by

$$i_t = A_1 \, e^{\delta_1 t} + A_2 \, e^{\delta_2 t}. \tag{1.7}$$

The general solution is then given by

$$i = i_s + i_t = \frac{\sqrt{2}|V|}{c - a\omega^2 + j\omega b} \, e^{j(\omega t + \phi)} + A_1 \, e^{\delta_1 t} + A_2 \, e^{\delta_2 t}. \tag{1.8}$$

Here A_1 and A_2 are arbitrary constants, which are to be determined from the initial conditions. All terms in equation (1.8), which include both steady- and transient-state solutions, are spiral vectors, the steady-state solution i_s being a circular vector, which is a spiral vector with $\lambda = 0$. Figure 1.2 shows the spiral vectors in equation (1.8) for $t = 0$, in the complex plane. The figure corresponds to the initial condition of zero currents at $t = 0$. By putting $t = 0$, both steady and transient vectors can be depicted as stationary vectors in the complex plane.

Thus, when state variables are expressed as spiral vectors, both steady and transient states can be treated in an unified way. For steady-state solutions $d/dt = p$ becomes $j\omega$, and for transient solutions p becomes δ. For integration, $1/p$ becomes $1/j\omega$ and $1/\delta$ respectively.

Spiral vector analysis of electrical circuits will now be treated more generally. For the lumped constant circuit, the general circuit equation can be written as follows:

$$i = \frac{A(p)}{B(p)} \, v. \tag{1.9}$$

Here, $A(p)$ and $B(p)$ are polynomials in p. When the input voltage v is given by the circular vector of equation (1.3), the steady-state solution of equation

(1.9) is obtained by inserting $p = j\omega$, as follows:

$$i_s = \frac{A(j\omega)}{B(j\omega)} \sqrt{2} V e^{j\omega t} = \frac{A(j\omega)}{B(j\omega)} \sqrt{2} \dot{V}. \tag{1.10}$$

Here, $B(j\omega)/A(j\omega)$ is the AC impedance at the input terminals. The characteristic equation is given by

$$B(p) = 0. \tag{1.11}$$

Let the characteristic roots of equation (1.11) be denoted $\delta_1, \delta_2, \ldots, \delta_n$. Then the general transient solution of equation (1.9) is given by

$$i_t = A_1 e^{\delta_1 t} + A_2 e^{\delta_2 t} + \cdots + A_n e^{\delta_n t}. \tag{1.12}$$

Here, $A_1, A_2, \ldots, A_n$ are arbitrary constants, which are generally complex and are to be determined by initial conditions. If δ_r is the mth-multiple root among the characteristic roots, the following terms must be included in i_t of equation (1.12)

$$A_1' t e^{\delta_r t} + A_2' t^2 e^{\delta_r t} + \cdots + A_{m-1}' t^{m-1} e^{\delta_r t}. \tag{1.13}$$

But the following discussion is not influenced by the inclusion of these terms.

The general solution is now given by

$$i = i_s + i_t = \frac{A(j\omega)}{B(j\omega)} \sqrt{2} |V| e^{j\omega t} + A_1 e^{\delta_1 t} + A_2 e^{\delta_2 t} + \cdots + A_n e^{\delta_n t}. \tag{1.14}$$

Here, all terms of the solution are spiral vectors. This general solution shows that the spiral vector is the most natural form of variable expression for writing circuit equations and their solutions.

When the coefficients of the characteristic equation for $B(p)$ (equation (1.11)) are all real, the characteristic roots, when complex, become conjugate pairs. However, when the coefficients are not all real, the characteristic roots are not conjugate pairs.

When the forcing function v of equation (1.9) is a spiral vector given by

$$v = \sqrt{2} |\dot{V}| e^{\delta t}, \qquad \delta = -\lambda + j\omega, \tag{1.15}$$

the general solution of equation (1.9) becomes

$$i = \frac{A(\delta)}{B(\delta)} v + A_1 e^{\delta_1 t} + A_2 e^{\delta_2 t} + \cdots + A_n e^{\delta_n t}. \tag{1.16}$$

This corresponds to the general solution of equation (1.14), which is valid for the forcing function of a circular vector. Here, $j\omega$ in the first term of equation (1.14), which is the steady-state term, is replaced by δ, which is complex. The first term is not a steady-state term but is a spiral vector. Let us call this the quasi-steady-state term, and call the circuit the spiral vector AC circuit. The circuit may be called the circular vector AC circuit when its forcing function is a circular vector.

Equations (1.15) and (1.16) open up a new area of the electric circuit theory and its practical applications. (Refer to equation (V.38) in Appendix V.)

1.2 Spiral vectors and corresponding real values

When i represents a current, it can be expressed in various ways, and a suitable expression must be chosen for our purposes. For example, when i is a cosine function of time, we can write

$$i = \sqrt{2}|I| \cos(\omega t + \phi), \tag{1.17}$$

which can also be expressed as

$$i = \mathrm{Re}\left[\sqrt{2}|I|\, e^{j(\omega t + \phi)}\right] = \mathrm{Re}\left[\sqrt{2}I\, e^{j\omega t}\right] = \mathrm{Re}\left[\sqrt{2}\dot{i}\right], \tag{1.18}$$

where $\mathrm{Re}[z]$ denotes the real part of z, and $I = |I|\, e^{j\phi}$ is the phasor of i. We can also represent i by the following spiral vector:

$$i = \sqrt{2}|I|\, e^{j(\omega t + \phi)} = \sqrt{2}\dot{i}. \tag{1.19}$$

Let us assume that in spiral vector theory a small letter indicates a spiral vector. Then the corresponding instantaneous real value i_{re} is given by

$$i_{re} = \mathrm{Re}[i]. \tag{1.20}$$

In the following, unless otherwise stated, state variables are all represented by spiral vectors whose symbols are small letters. Thus, we have

$$i = A\, e^{\delta t}, \qquad \delta = -\lambda + j\omega. \tag{1.21}$$

When $\lambda = 0$, the spiral vector becomes the vector of equation (1.19), which depicts a circle in the complex plane and represents steady-state AC current. Let us introduce the circular vector as follows:

$$\dot{I} = |I|\, e^{j(\omega t + \phi)} = I\, e^{j\omega t}. \tag{1.22}$$

Here, $I = |I|\, e^{j\phi}$ is a phasor, and we have the following relation:

$$|I| = |\dot{I}|. \tag{1.23}$$

When $\omega = 0$, i of equation (1.21) represents decaying DC. When $\delta = 0$, that is, $\lambda = 0$ and $\omega = 0$, i represents steady-state DC. When $t = 0$, $\dot{I}$ of equation (1.22) represents I, the phasor. Thus the spiral vector can represent almost all kinds of state variables that appear in electrical engineering.

1.3 Determination of spiral vectors

For the spiral vector given by equation (1.21), δ is the characteristic root of the circuit concerned, just as ω is the angular frequency of the power supply.

In equation (1.21) $A = |A| e^{j\phi_A}$ is a complex constant, which has two parameters, $|A|$ and ϕ_A. Two conditions are required to determine them. For example, if two initial conditions $i = i_0$ and $di/dt = i'_0$ at $t = 0$, which are real values, are given, then from equations (1.20) and (1.21) we get

$$i_0 = |A| \cos \phi_A, \tag{1.24}$$

$$i'_0 = -|A|(\lambda \cos \phi_A + \omega \sin \phi_A), \tag{1.25}$$

whence we get

$$\tan \phi_A = -(i'_0 + \lambda i_0)/\omega i_0, \tag{1.26}$$

$$|A| = \{i_0^2 + [\tfrac{1}{3}(i'_0 + \lambda i_0)]^2\}^{\frac{1}{2}}. \tag{1.27}$$

Thus $A = |A| e^{j\phi_A}$ has been determined. This may seem to be a lengthy process. In phasor notation, however, the current phasor $I = |I| e^{j\phi}$ requires two initial conditions and about the same amount of processing as equations (1.24)–(1.27) for the determination of $|I|$ and ϕ. As this process is usually omitted in the phasor method, the above process will also be omitted in the spiral vector method. After we get used to the method, the spiral vector representation is more convenient and easier to handle than the instantaneous real value representation.

1.4   Unified theory of AC circuits

The spiral vector can now be used to analyse AC circuits in an unified way, where both steady- and transient-state variables are expressed as spiral vectors. Steady-state analysis of AC circuits has been very well performed in the past by using phasor notation. But transient-state analysis of AC circuits has not been done well because the instantaneous real value expression has been used. Analysis of both states would now be better done in terms of spiral vectors.

State variables are time functions and spiral vectors are also time functions. The spiral vector expression for steady-state AC current is

$$i = \sqrt{2}|I| e^{j(\omega t + \phi)} = \sqrt{2I} e^{j\omega t} = \sqrt{2}\dot{I}. \tag{1.28}$$

Here, $\dot{I}$ is the circular vector introduced in equation (1.22). With the circular vector, we get the mathematical relations

$$\frac{d\dot{I}}{dt} = p\dot{I} = j\omega\dot{I} \quad \text{and} \quad \int \dot{I} \, dt = \frac{\dot{I}}{p} = \frac{\dot{I}}{j\omega}, \tag{1.29}$$

which put at our disposal the convention of $p = j\omega$ for steady-state AC analysis. In many books on AC circuit theory, authors have difficulty in explaining the conventions in equation (1.29), because the phasor is not a time function.

It is said that a merit of the phasor notation is the ease of presenting it as a stationary vector on the complex plane. When time is set to zero, the circular vector $\dot{i}$ in equation (1.28) becomes identical to the corresponding phasor, and it can be depicted as a stationary vector in the complex plane. It should be added here that the spiral vector representing a transient state can also be depicted as a stationary vector in the complex plane if time is set equal to zero. Therefore the wealth of works on steady-state AC circuits and machines written in terms of phasors remain relevant for spiral vector theory. The greatest merit of spiral vector theory is in the transient analysis of AC circuits and machines, where the phasor method is not applicable.

Appendix I gives an example of spiral vector analysis for a stationary AC circuit, showing that steady-state and transient-state solutions are simultaneously obtained in an unified way, and also showing that the transient-state solution is more easily obtained in terms of spiral vectors than in real values.

This book will not go further into detailed explanations of spiral vector theory for stationary AC circuits, but the author strongly recommends the replacement of the phasor notation and the real value expression by the spiral vector. He believes that it will also bring about good results in teaching AC circuits.

The spiral vector method exhibits its greatest analytical power in analyzing transient phenomena in three-phase AC machines, a topic which has been left largely uninvestigated. This vacant area of transient analysis of AC machines will be treated entirely by the spiral vector method in the second part of the book, and we will give many examples of spiral vector analysis.

PART II
AC motor analysis and control

2 Classical and modern motor control

Almost all types of electric motors were invented before the end of the nineteenth century, and their practical use also began in that century. They have been used increasingly widely, and now they are the main driving machines in industry, transportation, homes and many other fields. There are many kinds of both DC and AC motors. They have various characteristics and performances and the most suitable motors can be chosen for various applications. High-performance elevators and steel mills, for example, use separately excited DC motors, and railway trains are driven by DC series motors. Pumps and blowers are driven by three-phase induction motors. In these drives, steady-state torque–speed characteristics of the motor and the driven machine must be well matched, and controls in these applications have been mostly steady-state speed controls.

These problems have been treated by electric motor applications engineering, where steady-state performances have been the main concerns. And even in the control of start and stop and speed, speed adjustment under steady-state conditions was the main concern. Electromagnetic transient phenomena between successive steady states have not been taken into serious consideration. Even when transients are considered, they are usually dynamic transients of the equipment driven by the motor whose electromagnetic transients were not considered. The dynamic time constants of the driven equipment are usually much larger than those of the electromagnetic transients of the motor, the former time constants being dominant in the whole control system.

In motor applications engineering where speed control of the dynamic system was the main concern, control was classical motor control, where electromagnetic transients were unimportant and consequently received little attention.

2.1 Modern motor control

Control theory and techniques made remarkable progress during the Second World War, and knowledge in these fields has continued to increase.

Feedback control theory and techniques were established and introduced into industry, transportation, and our everyday lives, bringing about tremendous changes in all aspects of the post-war world. The electric motors placed in feedback loops were mostly DC control motors, called servomotors, because the controllability of AC motors was much inferior to that of DC motors.

Advances in power electronics initiated by semiconductor devices greatly increased the controllability of power supplies and led to progress in motor control technology. In classical motor control, the variable voltage power supply usually consisted of a motor-generator or a mercury arc rectifier. At this time, the controllable power supply was DC, while AC power supplies were difficult to control. Control motors were therefore usually DC motors. Power electronics first strengthened the utility of the DC motor as a control motor. This was because, as DC power supplies shifted from motor-generators and mercury arc rectifiers to semiconductor converters, AC to DC power conversion became easy and economical. Thus DC control motors, servomotors, flourished throughout industry, transportation, and automation. This was the beginning of modern motor control.

One of the characteristic features of modern motor control is that the controlled output is motor torque. The motor is included in the closed loop of modern control, and, irrespective of whether the control objective is speed, position, or some other physical quantities, the motor situated in the forward path of the loop is controlled with respect to its shaft torque. In classical motor control, the control objective was usually motor speed, while in the modern motor control the control objective is motor torque.

Control input to the motor is voltage or current at the input terminals, and the response of the motor torque becomes important. Electromagnetic transients within the motor thus become our main concern. In order to obtain a fast torque response, the electromagnetic transient must be analysed and means to suppress it must be found.

Another aspect of the modern motor control is the shift from DC motors to AC motors. Before the recent development of power electronics, it was difficult to change the frequency of the AC power supply. AC motors were tied to the synchronous speed and this was not easy to control. Progress in power electronics has freed AC motors from a fixed synchronous speed and has thus improved AC motor controllability.

One problem in modern motor control is the analytical theory of AC motors. Transient phenomena of AC motors have not been adequately and sufficiently analysed, owing to the lack of appropriate theories. Adequate analytical solutions have not been obtained by the conventional theories, such as the equivalent two-phase machine theory, the d–q axis theory, or the space vector theory,[19] etc. AC motor control, usually speed control, is based mainly on numerical solutions provided by computers. Thus its theoretical basis has been very weak.

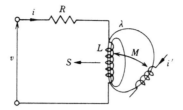

Fig. 2.1   Electrical circuit containing resistance and inductance

Spiral vector theory, briefly explained in the preceding chapter, will be now applied to AC motor analysis. Its analytical power is so strong that it reveals many useful characteristics and features of AC motors that were unknown. It will be shown that control features of AC motors are far superior to those of DC motors.

2.2   Fundamental equations for AC motors

AC motors have many windings. In three-phase motors, the primary side has three phase windings and the secondary side usually also has at least three phase windings. These windings usually have resistances and inductances as their circuit constants, as shown in Fig. 2.1. The circuit equation can be generally written as follows:

$$v = Ri + \frac{d\lambda}{dt}. \tag{2.1}$$

Here, λ is the flux linkage and is given by

$$\lambda = Li, \tag{2.2}$$

where L is the inductance. Inserting equation (2.2) in equation (2.1), we get

$$v = Ri + \frac{d}{dt}(Li) = Ri + L\frac{di}{dt} + i\frac{dL}{dx}\frac{dx}{dt}$$

$$= Ri + L\frac{di}{dt} + i\frac{dL}{dx}S. \tag{2.3}$$

Here, the third term is the speed voltage, which is proportional to the speed S of the winding along the coordinate x. When there is another winding or coil in the vicinity in which current i' flows and it is electromagnetically coupled, let the mutual inductance be denoted by M. Then λ in equation

(2.2) becomes

$$\lambda = Li + Mi', \tag{2.4}$$

and equation (2.3) becomes

$$v = Ri + \frac{d}{dt}(Li) + \frac{d}{dt}(Mi'). \tag{2.5}$$

When there are more windings, more terms corresponding to the third term in equation (2.5) are added to the equation. Equation (2.1) is the fundamental equation of the electric machine and is valid for any kind of variable expression, including the spiral vector.

3 Spiral vector analysis of the induction motor

3.1 Circuit equation of the induction motor: phase segregation method

Figure 3.1 shows a model for the induction motor analyses in this book. A three-phase induction motor is presented, and, unless otherwise specified, this is the motor assumed throughout this book. On the primary or stator side there are three phase windings, and on the secondary or rotor side there are also three phase windings. Although they are drawn as concentrated windings, their coils are inserted into slots pheripherally distributed on the cylindrical surface of the stator and rotor. When each phase winding is the same in structure and when the three phase windings are symmetrically positioned in space, the motor is called a symmetrical motor. The following analyses will treat only those symmetrical machines for which the three phase windings have the same circuit constants.

When the motor has a cage winding, the number of the secondary phases is much more than three. But it can be represented by the equivalent three-phase wound-rotor of Fig. 3.1.

Let the self-inductance of the primary and secondary windings be respectively denoted by L_1 and L_2. These can be divided into two parts:

$$L_1 = l_1 + M, \qquad L_2 = l_2 + M. \tag{3.1}$$

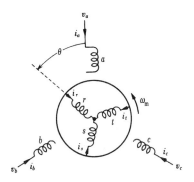

Fig. 3.1 Induction motor model

Here, l_1 and l_2 are the primary and secondary leakage inductances, and M is the maximum value of the mutual inductance between the primary and secondary phase windings. As shown in Fig. 3.1, if the spatial phase difference between phase a and phase r is θ, then the mutual inductance between the two phases is

$$M \cos \theta, \tag{3.2}$$

which assumes sinusoidal distribution of the magnetic field in the air gap. Here, θ is expressed in electrical radians (erad), which are related to mechanical radians by

$$\theta = \tfrac{1}{2}P\theta_{\text{mech}} \qquad (P: \text{number of poles}). \tag{3.3}$$

Equation (2.1) gives the following circuit equation for phase a:

$$v_a = R_1 i_a + \frac{d\lambda_a}{dt}. \tag{3.4}$$

The flux linkage λ_a can be divided into two parts:

$$\lambda_a = l_1 i_a + \lambda_{ga}. \tag{3.5}$$

The first term is the primary leakage flux linkage and λ_{ga} is the flux linkage of the revolving field in the air gap, which is also called main flux. Inserting equation (3.5) in equation (3.4), we get

$$v_a = R_1 i_a + l_1 p i_a + p\lambda_{ga}. \tag{3.6}$$

For phase r whose terminals are short-circuited, the circuit equation is

$$0 = R_2 i_r + l_2 p i_r + p\lambda_{gr}. \tag{3.7}$$

The main flux linkages λ_{ga} and λ_{gr} are produced by primary three-phase and secondary three-phase currents and are given by

$$
\begin{aligned}
\lambda_{ga} &= M i_a + M i_b \cos \tfrac{2}{3}\pi + M i_c \cos(-\tfrac{2}{3}\pi) \\
&\quad + M i_r \cos \theta + M i_s \cos(\theta + \tfrac{2}{3}\pi) + M i_t \cos(\theta - \tfrac{2}{3}\pi) \\
&= M[i_a - \tfrac{1}{2}(i_b + i_c)] + M\{[i_r - \tfrac{1}{2}(i_s + i_t)]\cos\theta - \tfrac{1}{2}\sqrt{3}(i_s - i_t)\sin\theta\},
\end{aligned} \tag{3.8}
$$

$$
\begin{aligned}
\lambda_{gr} &= M i_r + M i_s \cos \tfrac{2}{3}\pi + M i_t \cos(-\tfrac{2}{3}\pi) \\
&\quad + M i_a \cos(-\theta) + M i_b \cos(-\theta + \tfrac{2}{3}\pi) + i_c \cos(-\theta - \tfrac{2}{3}\pi) \\
&= M[i_r - \tfrac{1}{2}(i_s + i_t)] + M\{[i_a - \tfrac{1}{2}(i_b + i_c)]\cos\theta + \tfrac{1}{2}\sqrt{3}(i_b - i_c)\sin\theta\},
\end{aligned} \tag{3.9}
$$

Under steady-state operation, three-phase currents are symmetrical and can be represented by the following spiral vectors, which are in this case circular vectors:

$$i_a = \sqrt{2}|\dot{I}_1|\, e^{j(\omega_1 t + \phi_1)}, \quad i_b = \sqrt{2}|\dot{I}_1|\, e^{j(\omega_1 t + \phi_1 - \frac{2}{3}\pi)}, \quad i_c = \sqrt{2}|\dot{I}_1|\, e^{j(\omega_1 t + \phi_1 + \frac{2}{3}\pi)},$$

(3.10)

$$i_r = \sqrt{2}|\dot{I}_2|\, e^{j(\omega_2 t + \phi_2)}, \quad i_s = \sqrt{2}|\dot{I}_2|\, e^{j(\omega_2 t + \phi_2 - \frac{2}{3}\pi)}, \quad i_t = \sqrt{2}|\dot{I}_2|\, e^{j(\omega_2 t + \phi_2 + \frac{2}{3}\pi)}.$$

(3.11)

Here, ω_1 and ω_2 are the angular frequencies of the primary and secondary currents, respectively, and are different from each other, because they are independently observed on the stator side and the rotor side. Since the three-phase currents in equations (3.10) and (3.11) are symmetrical spiral vectors, we have

$$i_a + i_b + i_c = 0, \qquad i_r + i_s + i_t = 0 \tag{3.12}$$

and

$$i_b - i_c = -j\sqrt{3}i_a, \tag{3.13}$$

$$i_s - i_t = -j\sqrt{3}i_r. \tag{3.14}$$

Inserting equations (3.12)–(3.14) in equations (3.8) and (3.9), we have

$$\lambda_{ga} = \tfrac{3}{2}Mi_a + \tfrac{3}{2}Mi_r\, e^{j\theta}, \tag{3.15}$$

$$\lambda_{gr} = \tfrac{3}{2}Mi_r + \tfrac{3}{2}Mi_a\, e^{-j\theta}. \tag{3.16}$$

And inserting equations (3.15) and (3.16) in equations (3.6) and (3.7), we have

$$v_a = R_1 i_a + l_1 p i_a + \tfrac{3}{2}M[p i_a + p(i_r\, e^{j\theta})], \tag{3.17}$$

$$0 = R_2 i_r + l_2 p i_r + \tfrac{3}{2}M[p i_r + (p i_a)\, e^{-j\theta} - j\omega_m i_a\, e^{-j\theta}]. \tag{3.18}$$

Here, ω_m is the motor's angular velocity, given by

$$\omega_m = \frac{d\theta}{dt} = p\theta \qquad [\text{erad s}^{-1}]. \tag{3.19}$$

Multiplying equation (3.18) by $e^{j\theta}$, we get

$$0 = R_2 i_r\, e^{j\theta} + (l_2 + \tfrac{3}{2}M)(p i_r)\, e^{j\theta} + \tfrac{3}{2}M(p - j\omega_m)i_a. \tag{3.20}$$

Making the variable replacement

$$i_r' = i_r\, e^{j\theta} = i_r\, e^{j\omega_m t}, \tag{3.21}$$

we get

$$p i_r' = (p i_r)\, e^{j\theta} + j\omega_m i_r'. \tag{3.22}$$

Inserting equations (3.21) and (3.22) in equations (3.17) and (3.20), we get

$$v_a = R_1 i_a + (l_1 + \tfrac{3}{2}M)p i_a + \tfrac{3}{2}M p i_r', \tag{3.23}$$

$$0 = R_2 i_r' + (l_2 + \tfrac{3}{2}M)(p - j\omega_m)i_r' + \tfrac{3}{2}M(p - j\omega_m)i_a. \tag{3.24}$$

It should be noticed here that the voltage and currents in these two equations are of phases a and r only, the other phases of the primary and secondary three phases being left out. Phases a and r are thus segregated from the other phases. This is called phase segregation. This means that one only of the three phases is sufficient to write the circuit equation of the three-phase motor.

In spite of equations (3.10) and (3.11), which assume steady-state operation, equations (3.23) and (3.24) are valid for both steady- and transient-state operations, as will be shown in Chapter 4.

In order to let phase a represent the primary and phase r represent the secondary, subscripts a and r are changed to 1 and 2 respectively. Then equations (3.23) and (3.24) become

$$v_1 = R_1 i_1 + (l_1 + \tfrac{3}{2}M)p i_1 + \tfrac{3}{2}M p i_2, \tag{3.25}$$

$$0 = R_2 i_2 + (l_2 + \tfrac{3}{2}M)(p - j\omega_m)i_2 + \tfrac{3}{2}M(p - j\omega_m)i_1. \tag{3.26}$$

Here, $i_2 = i'_r$ of equation (3.21). From equation (3.21), the secondary frequency now becomes $\omega_m + \omega_2 = \omega_1$, which is equal to the power supply frequency ω. The time derivative p of the current becomes $j\omega$. Therefore equations (3.25) and (3.26) can now be written as

$$v_1 = R_1 i_1 + j\omega(l_1 + \tfrac{3}{2}M)i_1 + \tfrac{3}{2}j\omega M i_2, \tag{3.27}$$

$$0 = \tfrac{3}{2}j\omega M i_1 + \frac{R_2}{s} i_2 + j\omega(l_2 + \tfrac{3}{2}M)i_2, \tag{3.28}$$

where s is the slip, given by

$$s = \frac{\omega - \omega_m}{\omega}. \tag{3.29}$$

Equations (3.27) and (3.28) are the circuit equations of the induction motor under steady-state operation. They are valid only for state variables expressed as spiral vectors, which in this case are all circular vectors. Replacing the spiral vectors with the corresponding circular vectors, equations (3.27) and (3.28) become

$$\dot{V}_1 = R_1 \dot{I}_1 + j(x_1 + x_m)\dot{I}_1 + jx_m \dot{I}_2, \tag{3.30}$$

$$0 = jx_m \dot{I}_1 + \frac{R_2}{s} \dot{I}_2 + j(x_2 + x_m)\dot{I}_2. \tag{3.31}$$

The reactances here are

$$x_1 = \omega l_1, \qquad x_2 = \omega l_2, \qquad x_m = \tfrac{3}{2}\omega M, \tag{3.32}$$

where x_1 is the primary leakage reactance, x_2 is the secondary leakage

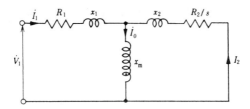

Fig. 3.2 T type steady-state equivalent circuit of the induction motor

reactance, and x_m is the exciting reactance. These equations appear to be the same as the performance equations expressed in phasors. Thus the spiral vector method is thoroughly compatible with the phasor method, although it should be remembered that the spiral vector (or circular vector) is a time function, while the phasor is not.

Figure 3.2 shows the T type steady-state equivalent circuit for the circular vector expression corresponding to equations (3.30) and (3.31). This is the same as the well-known T type equivalent circuit for the phasor expression. Even in today's computer age, the equivalent circuit is very useful for visualizing electromagnetic phenomena occurring within the induction motor. It first reveals the existence of the exciting current $\dot{I}_0$, which is given by

$$\dot{I}_0 = \dot{I}_1 + \dot{I}_2. \tag{3.33}$$

As will be shown throughout this book, the exciting current plays a very important role in the analysis and control of the induction motor.

In the equivalent circuit of Fig. 3.2, there is no resistance in parallel with the exciting reactance x_m. This is due to the neglect of the no-load iron loss. The error due to this neglect is very small, however, and, if necessary, a resistance representing the no-load iron loss can be inserted at the input terminals.

3.2   Steady-state analysis of the induction motor

In the preceding section, the circuit equation and T type equivalent circuit were derived for steady-state operation of the induction motor. Their mathematically rigorous derivation, as here, has rarely been done before.

The steady-state characteristics of the induction motor have been well investigated. The well-known parts of the analysis will not be repeated here, but their summary will be given. The T type equivalent circuit gives the following equations for voltage and currents in the induction motor:

$$\dot{I}_1 = \frac{\dot{V}_1}{R_1 + \dfrac{jx_m(R_2/s + jx_2)}{R_2/s + j(x_m + x_2)}}, \tag{3.34}$$

$$\dot{I}_2 = -\frac{jx_m}{R_2/s + j(x_m + x_2)}, \tag{3.35}$$

$$\dot{I}_0 = \frac{R_2/s + jx_2}{R_2/s + j(x_m + x_2)}\dot{I}_1, \tag{3.36}$$

$$\dot{I}_0 = \dot{I}_1 + \dot{I}_2. \tag{3.37}$$

The secondary copper loss is

$$P_c = 3R_2|\dot{I}_2|^2. \tag{3.38}$$

The secondary input carried through the air gap to the secondary circuit is

$$P_2 = 3\frac{R_2}{s}|\dot{I}_2|^2. \tag{3.39}$$

From equations (3.38) and (3.39), we get

$$P_c = sP_2. \tag{3.40}$$

The motor output is therefore

$$P_0 = P_2 - P_c = 3R_2\frac{1-s}{s}|\dot{I}_2|^2. \tag{3.42}$$

The motor's angular velocity is

$$\omega_m = \frac{2(1-s)}{P}\omega \quad [\text{rad s}^{-1}]. \tag{3.43}$$

Here, P is the number of poles and ω is the angular frequency of the power supply in electrical radian/second. The synchronous angular speed is given by

$$\omega_{syn} = \omega \quad [\text{erad s}^{-1}] = \frac{2\omega}{P} \quad [\text{rad s}^{-1}]. \tag{3.44}$$

The motor torque is then given by

$$T = \frac{P_0}{\omega_m} = \tfrac{3}{2}P\frac{R_2}{s\omega}|\dot{I}_2|^2 \quad [\text{N m}] = 3\frac{R_2}{s}|\dot{I}_2|^2 \quad [\text{syn W}] = P_2[\text{syn W}]. \tag{3.45}$$

The motor torque expressed in synchronous watts [syn W] is equal to the output produced by the motor if it would run at synchronous speed, and is equal to the secondary input power.

Table 3.1   Rating and circuit constants of an induction
motor

Rating	3.7 kW, 200 V, 50 Hz, 4 poles
Circuit	$x_1 = 0.4087\ \Omega$, $x_2 = 0.269\ \Omega$, $x_m = 13.2\ \Omega$
constants	$R_1 = 0.463\ \Omega$, $R_2 = 0.432\ \Omega$

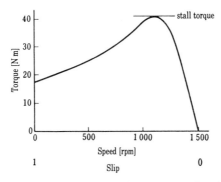

Fig. 3.3   Torque–speed curve of the induction motor of Table 3.1 for constant voltage and constant frequency

When the voltage and frequency of the power supply are kept constant, the currents and torque are functions only of slip, as indicated by the equivalent circuit in Fig. 3.2, where the variable circuit constant is R_2/s only.

Table 3.1 gives the rated values and circuit constants of an induction motor. Its torque–speed characteristics were calculated using equations (3.34)–(3.45) and are shown by the curve in Fig. 3.3. The torque is zero at the synchronous speed and it increases proportionally with increasing slip. But this linearity is soon lost and the torque reaches its maximum, which is called the stall torque. The corresponding slip is usually 20–30%. The rated torque must be well below the stall torque, and therefore slip at the rated torque is usually 3–5%. Thus, under constant-voltage and constant-frequency operation, the maximum torque is usually less than twice the rated torque, the overload torque capacity being very limited. The straight portion of the characteristic curve is also very short and the normal operation speed range is limited to this small range. Thus the induction motor has been considered to be practically a constant-speed motor with a poor control motor capacity. This was due, however, to the lack of an AC power supply frequency control, and to the inappropriate ways that were therefore used to control AC motors.

3.3   Scalar control of the induction motor

Before the advent of semiconductor power elements like power transistors and thyristors, it was not easy to control the voltage, current, or frequency of AC power supplies. Although power supply control must precede motor control, it was difficult to control the output of AC power supplies as vector quantities, and the voltage, current, and frequency were controlled as scalar quantities. Thus the control in classical AC motor control was scalar control, and fast response was not and could not be required. This was adjustment not control. DC motors were usually used when a fast control response was required.

The main concern of classical AC motor control was speed control, which was scalar control. Scalar control of the induction motor has been well treated and therefore only its main points will be summarized here. In the voltage control of the motor speed, the magnitude of the primary voltage $|V_1|$ is controlled in order to control the motor speed. Since torque is proportional to $|V_1|^2$, the torque–speed curves change as shown in Fig. 3.4. Motor speeds, which are determined as intersections of the motor and load torque–speed curves, decrease as voltage decreases. However, as the stall torque also decreases, the control range of speed is very limited. No-load speed cannot be controlled. Thus speed control with a variable-voltage constant-frequency power supply is very limited.

When the power supply frequency is controlled, as shown in Fig. 3.5, where $|V_1|$ is also controlled to keep $|V_1|/f$ constant. The stall torque then remains almost unchanged, and the controllable speed range becomes much larger. Thus the variable-voltage variable-frequency power supply gives better speed control for the induction motor. But until recently frequency control was neither technically easy nor economical.

In the wound rotor type of induction motor, it is possible to change the secondary resistance R_2, and for different values of R_2 the torque–speed

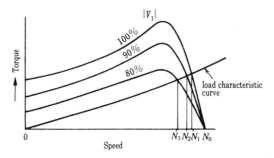

Fig. 3.4   Voltage control of induction motor speed

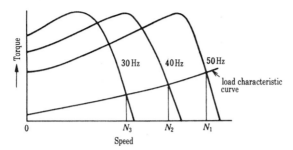

Fig. 3.5   Frequency control of induction motor speed for constant V/f

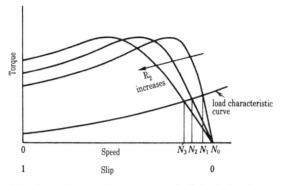

Fig. 3.6   Secondary resistance control of the induction motor

curves are shifted to the left with R_2/s remaining constant, as shown in Fig. 3.6. This is called proportional shifting of the torque–speed curves. As secondary resistance increases, motor speed decreases. However, the no-load speed remains unchanged and the secondary resistance loss increases at lower motor speed.

These motor speed controls are speed adjustments under steady-state electromagnetic conditions, where neither fast response nor precision control are required. Scalar control of AC motors was not motor control as in the servomotor. During the time when AC motors were scalar-controlled, they were much inferior to the DC servomotor.

3.4   Linearization of torque–speed characteristics of the induction motor

As shown in Figs 3.3–3.6, the torque–speed curves of the induction motor have rather short linear portions in the small slip range, and the stall torque is low, about twice the rated torque. This decreases overload torque capacity

and thus reduces control-motor performance quality. This inferior performance, however, comes from the constant-voltage constant-frequency (CVCF)
operation of the motor. This will be now improved by a better way of
operating the motor, which is the first step in the modern control of the
induction motor.

From equations (3.35) and (3.36), we get

$$\dot{I}_2 = -\frac{jx_m}{R_2/s + jx_2}\dot{I}_0. \tag{3.46}$$

Inserting this equation in equation (3.45) gives

$$T = \tfrac{3}{2}P\frac{R_2}{s\omega}\frac{(sx_m)^2}{R_2^2 + (sx_2)^2}|\dot{I}_0|^2 \quad [\text{N m}]. \tag{3.47}$$

When the exciting current $|\dot{I}_0|$ is kept constant, the torque T is a function
only of slip frequency (sf or $s\omega$). Then, for the motor in Table 3.1, Fig. 3.7
shows torque T in equation (3.47) as a function of motor speed for different
power supply frequencies. The torque–speed curves are straight up to six
times the rated torque, and the stall torque points do not even appear in
this figure. The torque–speed curve for a constant voltage of 200 V and a
constant frequency of 50 Hz is shown by the broken line. Compared with
the CVCF curve, constant exciting-current curves have very much improved

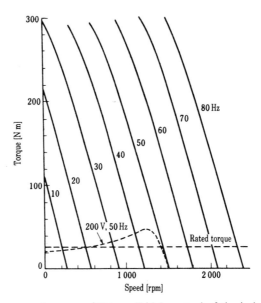

Fig. 3.7   Torque–speed curves of T type FAM control of the induction motor

linearity. A constant exciting current $|I_0|$ keeps the flux density of the revolving magnetic field in the air gap constant, and thus most parts of magnetic materials in the motor are utilized at a constant level. This motor control, which keeps the exciting-current amplitude constant, is called the field acceleration method (FAM).[1] FAM will be utilized for induction motor torque control throughout this book, and it will be shown that it derives its very superior control features from the induction motor.

3.5 Equivalent circuit transformation of the induction motor

The equivalent circuit of the induction motor is not unique. There are infinite numbers of them, and we must choose an appropriate one for our purposes. The infinite number of equivalent circuits are equivalent with respect to input impedance at the primary terminals. By transforming the T type equivalent circuit, various equivalent circuits will be derived.

The circuit equations (3.30) and (3.31) are combined into the following matrix equation:

$$\begin{bmatrix} \dot{V_1} \\ 0 \end{bmatrix} = \begin{bmatrix} R_1 + j(x_1 + x_m) & jx_m \\ jx_m & R_2/s + j(x_2 + x_m) \end{bmatrix} \begin{bmatrix} \dot{I_1} \\ \dot{I_2} \end{bmatrix}. \tag{3.48}$$

Here, the state variables are represented by circular vectors. Now currents are transformed by the transformation equation

$$\begin{bmatrix} \dot{I_1} \\ \dot{I_2} \end{bmatrix} = \begin{bmatrix} 1 & 0 \\ 0 & \alpha \end{bmatrix} \begin{bmatrix} \dot{I_1} \\ \dot{I_2} \end{bmatrix} = C \begin{bmatrix} \dot{I_1} \\ \dot{I_2^\alpha} \end{bmatrix}, \tag{3.49}$$

where α is an arbitrary constant. In this transformation, $\dot{I_1}$ remains unchanged and the secondary current I_2 is transformed to $I_2^\alpha = I_2/\alpha$. This means that the effective turns ratio is changed from 1 to α. Inserting equation (3.49) into equation (3.48) and multiplying both sides of equation (3.48) by C_t, we get

$$C_t[V] = C_t[Z]C[I], \tag{3.50}$$

This circuit transformation keeps the input impedance, power, and torque of the motor invariant, as will be shown below. If we perform the matrix multiplication, equation (3.50) gives

$$\begin{bmatrix} \dot{V_1} \\ 0 \end{bmatrix} = \begin{bmatrix} R_1 + j(x_1 + x_m) & j\alpha x_m \\ j\alpha x_m & \alpha^2 R_2/s + j\alpha^2(x_2 + x_m) \end{bmatrix} \begin{bmatrix} \dot{I_1} \\ \dot{I_2^\alpha} \end{bmatrix}. \tag{3.51}$$

Figure 3.8 shows the corresponding circuit. Its input impedance Z_1 at the

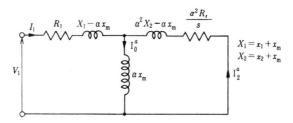

Fig. 3.8   General steady-state equivalent circuit of the induction motor

terminals is

$$Z_1 = \frac{\dot{V}_1}{\dot{I}_1}$$

$$= R_1 + j[x_1 - (1 - \alpha)x_m] + \left(\frac{1}{j\alpha x_m} + \frac{1}{\alpha^2 R_2/s + j[\alpha^2(x_2 + x_m) - \alpha x_m]}\right)^{-1}$$

$$= R_1 + jx_1 + \frac{jx_m(R_1/s + jx_2)}{R_2/s + j(x_2 + x_m)}, \tag{3.52}$$

and Z_1 is independent of α and equal to the input impedance of the T-type equivalent circuit in Fig. 3.2. The secondary input power is

$$P_2 = 3\frac{\alpha^2 R_2}{s}|I_2^a|^2 = 3\frac{R_2}{s}|I_2|^2, \tag{3.53}$$

which also remains unchanged. Accordingly, torque, which is proportional to P_2, is also invariant in this transformation. Since α is arbitrary, Fig. 3.8 indicates that there are an infinite number of equivalent circuits. Thus the circuit in Fig. 3.8 will be named the general steady-state equivalent circuit for the induction motor.

When $\alpha = 1$, the general equivalent circuit becomes the T type equivalent circuit in Fig. 3.2. We shall now derive two equivalent circuits from the general equivalent circuit, which are very useful in induction motor analysis and control.

If α is chosen as

$$\alpha = \frac{x_m}{x_m + x_2} = \frac{\frac{3}{2}M}{\frac{3}{2}M + l_2}, \tag{3.54}$$

then equation (3.51) becomes

$$\begin{bmatrix} \dot{V}_1 \\ 0 \end{bmatrix} = \begin{bmatrix} R_1 + j(x_1' + x_m') & jx_m' \\ jx_m' & R_2'/s + jx_m' \end{bmatrix} \begin{bmatrix} \dot{I}_1 \\ \dot{I}_2' \end{bmatrix}. \tag{3.55}$$

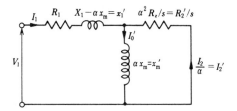

Fig. 3.9   T-I type steady-state equivalent circuit of the induction motor

Here, $R_2' = \alpha^2 R_2$, $x_1' = x_1 + x_m(1 - \alpha)$, $x_m' = \alpha x_m$, and $\dot{I}_2 = \dot{I}_2/\alpha$. Figure 3.9 shows the corresponding equivalent circuit, which is named the T-I type steady-state equivalent circuit. It should be noticed here that the secondary leakage reactance x_2' is zero.

If α is now chosen as

$$\alpha = \frac{x_m + x_1}{x_m} = \frac{\tfrac{3}{2}M + l_1}{\tfrac{3}{2}M},$$ (3.56)

then equation (3.51) becomes

$$\begin{bmatrix} \dot{V}_1 \\ 0 \end{bmatrix} = \begin{bmatrix} R_1 & jx_m'' \\ jx_m'' & R_2''/s + j(x_2'' + jx_m'') \end{bmatrix} \begin{bmatrix} \dot{I}_1 \\ \dot{I}_2'' \end{bmatrix}.$$ (3.57)

Here, $R_2'' = \alpha^2 R_2$, $x_m'' = \alpha x_m$, and $x_2'' = \alpha^2(x_2 + x_m) - \alpha x_m$. Figure 3.10 shows the corresponding equivalent circuit, which is named the T-II type steady-state equivalent circuit. It should be noticed here that the primary leakage reactance x_1'' is zero.

If the exciting reactance x_m'' is moved to the input terminals, the T-II type equivalent circuit in Fig. 3.10 becomes the L type steady-state equivalent circuit shown in Fig. 3.11. Its input impedance is not the same as that of the other types, and it has theoretical error. But its error is smaller than that of the conventional L type equivalent circuit for the induction motor, which is derived by moving the exciting reactance over $R_1 + jx_1$ to the terminals in the T type equivalent circuit.

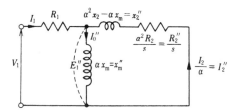

Fig. 3.10   T-II type steady-state equivalent circuit of the induction motor

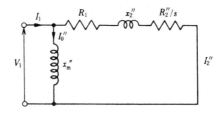

Fig. 3.11   L type steady-state equivalent circuit of the induction motor

3.6   Characteristic performance curves of the induction motor

Torque characteristic curves are most important for control motors or servomotors, and the control output in high-class motor control is motor torque. In the preceding section, four equivalent circuits, T type, T-I type, T-II type, and L type were derived. All, except the L type, have the same input impedance at the terminals. Three equivalent circuits produce the same output and torque for the same input voltage. The secondary side, however, including the exciting reactance, differs between the different equivalent circuits. Making use of the different secondary sides, we can obtain different control characteristics for the induction motor.

Figure 3.7 shows that the T type equivalent circuit produces torque–speed curves that are straight over very wide ranges for FAM motor control, which keeps the exciting current $|\dot{I}_0|$ constant.

In the T-I type equivalent circuit of Fig. 3.9, the following equation holds:

$$\dot{I}'_2 = -\mathrm{j}\,\frac{sX'_m}{R'_2}\,\dot{I}'_0. \tag{3.58}$$

The torque is given by

$$T = \tfrac{3}{2}P\,\frac{R'_2}{s\omega}|\dot{I}'_2|^2 = \tfrac{3}{2}P\,\frac{1}{R'_2}(\tfrac{3}{2}M')^2 s\omega|\dot{I}'_0|^2. \tag{3.59}$$

These equations indicate that the secondary current and torque are both proportional to the slip frequency, when the exciting current $|I'_0|$ is kept constant, that is,

$$T \propto |\dot{I}'_2| \propto sf. \tag{3.60}$$

From equations (3.58) and (3.59), we get

$$T = \tfrac{9}{4}M'P|\dot{I}'_0|\,|\dot{I}'_2|. \tag{3.61}$$

This makes the induction motor behave just like a DC motor. When the exciting current $|I'_0|$ is kept constant, as it is in FAM control, the torque T is proportional to secondary current $|\dot{I}'_2|$.

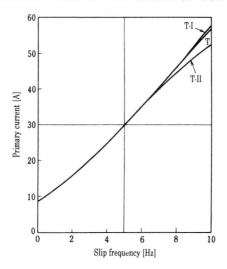

Fig. 3.12 Primary current versus slip frequency curves of the induction motor in Table 3.1

In the T-II type equivalent circuit of Fig. 3.10, the following equations hold:

$$\dot{I}_2'' = -j \frac{sx_m''}{R_2'' + jsx_2''} \dot{I}_0'',$$ (3.62)

$$T = \tfrac{3}{2}P \frac{R_2''}{s\omega} \frac{(sx_2'')^2}{R_2''^2 + (sx_2'')^2} |\dot{I}_2''|^2.$$ (3.63)

For the motor of Table 3.1, the exciting current is kept constant at the rated value of 8.7 A, and the primary current and torque were calculated. These are shown in Figs. 3.12 and 3.13 as functions of slip frequency. Three curves are based on the T, T-I, and T-II equivalent circuits. The T-I type has a completely straight torque–slip frequency curve. The other types also have good linearity in their torque–slip frequency curves. Because the rated slip frequency is 2.4 Hz, up to 2.5 times the rated torque the characteristic curves of the three types do not practically differ for currents or torque.

The T, T-I, and T-II type equivalent circuits are drawn in one circuit in Fig. 3.14. A corresponding vector diagram of the voltages and currents is shown in Fig. 3.15. This is the circular vector diagram for $t = 0$, which is identical to the phasor diagram. The slip frequency is chosen to be the same for the three equivalent circuits, so the exciting currents differ slightly between them.

The torque–speed curves for FAM control are obtained from Fig. 3.13.

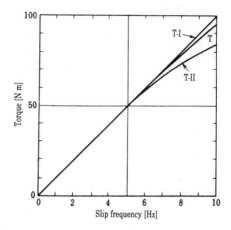

Fig. 3.13 Torque versus slip frequency curves of the induction motor in Table 3.1

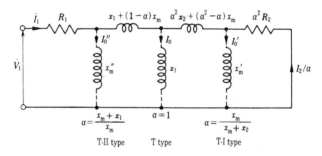

Fig. 3.14 T, T-I, and T-II type steady-state equivalent circuits of the induction motor

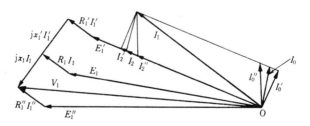

Fig. 3.15 Vector diagram of the induction motor for the same I_1

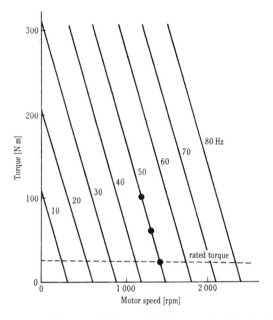

Fig. 3.16 Torque–speed curves of T-I type FAM control of the induction motor of Table 3.1 (· indicates measured torque)

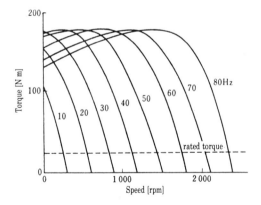

Fig. 3.17 Torque–speed curves of T-II type FAM control of the induction motor of Table 3.1

These are shown in Fig. 3.16 and 3.17 for the T-I and T-II type equivalent circuits. Power supply frequencies are attached to each curve as parameters. T-I type FAM control gives the perfectly straight torque–speed curves of Fig. 3.16. Three points measured for 50 Hz are on the straight line. This

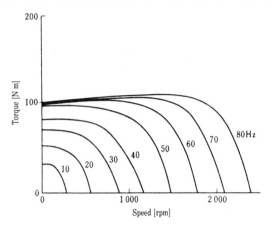

Fig. 3.18 Torque–speed curves of L type FAM control of the induction motor of Table 3.1

indicates that FAM control, where the exciting current $|\dot{I}'_0|$ is kept constant, practically eliminates the influence of iron saturation caused by the torque component current $|\dot{I}'_2|$.

The torque–speed curves of T-II type FAM control given in Fig. 3.17 are not perfectly straight, but their straight portions extend up to about three times the rated torque. All torque–speed curves are similar and parallel in both T-I and T-II type FAM controls as well as for the T type FAM control shown in Fig. 3.7, and the proportional shift principle holds.

Figure 3.18 shows torque–speed curves for the L type equivalent circuit. They are not similar or parallel and low-frequency curves have a very low stall torque. The L type equivalent circuit has very inferior control characteristics and should not be considered for practical application, and especially not for low-speed application.

3.7   FAM control of the induction motor torque

Induction motor torque can be controlled by controlling either the primary voltage or the primary current. These control modes are called respectively voltage-input control and current-input control. The appropriate one must first be chosen and then the appropriate equivalent circuit must be chosen for it.

The torque control explained here is FAM control, where the amplitude of the exciting current is kept constant. Then, for current-input control, the T-I type equivalent circuit of Fig. 3.9 is most appropriate. In this circuit, the proportional relation of equation (3.60) holds.

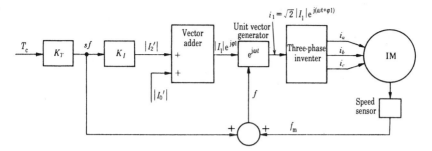

Fig. 3.19 Block diagram of T-I type FAM control (current-input control) of the induction motor

Using this relation, the block diagram of Fig. 3.19 shows how T-I type FAM control of induction motor torque is performed. The torque command T_c is given at the left end and then the coefficient multiplier K_T produces the slip frequency of sf. Another coefficient multiplier K_I determines $|I'_2|$, which is added perpendicularly to the exciting current $|I'_0|$, which is fixed at a rated value by the vector adder. This gives $|I_1| e^{j\phi} = I_1$. The power supply frequency is determined by adding sf and f_m (motor speed frequency). These are fed to a unit vector generator $e^{j\omega t}$ to obtain the instantaneous value i_1 of the primary current. This drives the three-phase inverter, to produce currents i_a, i_b, and i_c. And these are fed to the motor, to produce a torque equal to the torque command T_c. This block diagram uses only standard IC components and is simple in structure. It produces the straight torque–speed curves in Fig. 3.16. In this sense, T-I type FAM control produces perfect control characteristics. T-I type FAM control can also produce instantaneous control response in induction motor torque control, as will be explained in the next chapter.

For voltage-input control, the T-II type equivalent circuit of Fig. 3.10 is appropriate. In this circuit, the following equations hold:

$$\dot{V}_1 = R_1 \dot{I}_1 + \dot{E}''_1, \tag{3.64}$$

$$\dot{E}''_1 = jx''_m \dot{I}''_0 = \tfrac{3}{2} j\omega M'' \dot{I}''_0, \tag{3.65}$$

Here, E''_1 is the exciting voltage at the exciting reactance terminals. In FAM control, the exciting current $|\dot{I}''_0|$ is kept constant. And the block diagram in Fig. 3.20 then performs voltage-input FAM control. The torque command T_c is given at the left end and the slip frequency sf proportional to it is determined by coefficient multiplier K_T. Adding the motor speed frequency to it, the power supply frequency $f = sf + f_m$ is obtained. Feeding f and $|I''_0|$ to the unit vector generator $e^{j\omega t}$, the instantaneous value of the exciting voltage e''_1 is obtained. Adding $R_1 i_1$ to e''_1, the primary voltage v_1 is

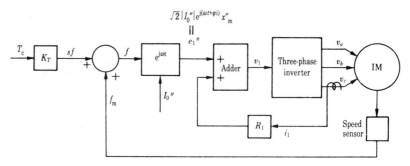

Fig. 3.20 Block diagram of T-II type FAM control (voltage-input control) of the induction motor

determined. The voltage signal v_1 drives the three-phase inverter, which generates three phase voltages v_a, v_b, and v_c. These are fed to the induction motor, to produce a torque equal to the torque command T_c. This block diagram is also simple in structure and uses only standard IC components. Steady-state torque–speed characteristic curves shown in Fig. 3.17 are obtained by T-II type FAM control and they are very good, though not perfectly straight. The transient response of this block diagram is not instantaneous, as is T-I control, but it can be very fast, as will be shown in the next chapter.

In the above explanation, the T-I type equivalent circuit was used for current-input control, while the T-II type equivalent circuit was used for voltage-input control. These combinations of the two equivalent circuits and the two control types are good, but other combinations are also possible. The T type equivalent circuit can also be used for both voltage- and current-input controls. This point will be explained in Chapter 9, where control of the three-phase inverter is explained.

4 Electromagnetic transient analysis of the induction motor

4.1 Introduction

As AC motor control began to shift from classical to modern control with the advent of power electronics, the lack of adequate analytical and control theories for AC motors was the first problem. This prevented full use of the favourable situation created by advances in power electronics. Without adequate theoretical analysis of electromagnetic transient phenomena in AC motors, the inherently superior features of AC motors could not be exploited. In analyzing AC motor transients, the equivalent two-phase machine theory, or two-axis theory (sometimes mistakenly called d–q axis theory), is popular. This theory, however, has not provided an adequate anaytical solution of AC motor transients.[18,19,20] AC motors have more windings than DC motors and so their analysis is more complicated than that of DC motors. The equivalent two-phase method reduces the number of phases from three to two, but this is still too many for derivation of analytical solutions of the performance equations. When analysing transients of motor speed control, where electromagnetic transients within the motor and dynamic transients of the driven equipment were mixed together, computers have provided numerical solutions. The former transient was thought to be influenced by the latter. But as will be shown in this chapter, these two transients can be made independent of each other by FAM control of the induction motor.

It is necessary to establish a sound analytical theory for the AC motor to complete its electromagnetic transient analysis. For this purpose, the spiral vector method and the phase segregation method, developed and very effectively applied to steady-state analysis in the preceding chapter, will now be applied to transient analyses of AC motors.

4.2 Transient analysis of the induction motor by the spiral vector and phase segregation methods

In Chapter 3, the spiral vector and phase segregation methods made it possible to write steady-state performance equations with one-phase variables representing three phases. The phase segregation method has been used for a very long time in analysing steady-state performances of the three-phase machines, but has not been applied to the analysis of transient phenomena. The spiral vector method makes it possible to apply the phase segregation method to transient analyses of AC machines. In Section 3.1, the circuit equations of the induction motor for the steady state were derived in terms of circular vectors. Spiral vector representation must be used for transient analysis, but the analytical process is about the same. The induction motor model given in Fig. 3.1 will also be used here for transient analysis.

4.2.1   *Transient analysis of current-input control of the induction motor*

In induction motor applications, the inverter power supply is more often the voltage source type than the current source type. For torque control, however, current-input control has many advantages and the current source inverter may be used for it (see Chapter 9). The starting equation for current-input control is given by equation (3.7), which is rewritten here:

$$0 = R_2 i_r + l_2 p i_r + p\lambda_{gr}. \tag{4.1}$$

The primary three-phase currents are the control input, and they are the symmetrical three-phase currents given by the following equations:

$$\left.\begin{array}{l} i_a = \sqrt{2}|\dot{I}_1| \, \mathrm{e}^{\mathrm{j}(\omega t + \phi_1)} = \sqrt{2}\dot{I}_1, \\[4pt] i_b = \sqrt{2}|\dot{I}_1| \, \mathrm{e}^{\mathrm{j}(\omega t + \phi_1 - \frac{2}{3}\pi)} = \sqrt{2}\dot{I}_1 \, \mathrm{e}^{-\frac{2}{3}\mathrm{j}\pi}, \\[4pt] i_c = \sqrt{2}|\dot{I}_1| \, \mathrm{e}^{\mathrm{j}(\omega t + \phi_1 + \frac{2}{3}\pi)} = \sqrt{2}\dot{I}_1 \, \mathrm{e}^{+\frac{2}{3}\mathrm{j}\pi}. \end{array}\right\} \tag{4.2}$$

These are spiral vectors with $\lambda = 0$, or circular vectors (see equation (1.28)), and they are the same as equations (3.10). Secondary three-phase currents, though, contain transient components and cannot be expressed by circular vectors, as given by equations (3.11). But equations (4.2) are sufficient to derive the main flux linkage λ_{gr} given below, which is the same as equation (3.16):

$$\lambda_{gr} = \tfrac{3}{2}M(i_r + i_a \, \mathrm{e}^{-\mathrm{j}\theta}). \tag{4.3}$$

Inserting this equation in equation (4.1), we get

$$0 = R_2 i_r + (l_2 + \tfrac{3}{2}M)p i_r + \tfrac{3}{2}M p(i_a \, \mathrm{e}^{-\mathrm{j}\theta}). \tag{4.4}$$

Inserting i_a of equation (4.2) with $\theta = \omega_m t$ in equation (4.4), we get

$$\begin{aligned} 0 &= R_2 i_r + (l_2 + \tfrac{3}{2}M)p i_r + \tfrac{3}{2}M p\big(\sqrt{2}|\dot{I}_1| \, \mathrm{e}^{\mathrm{j}(\omega t - \omega_m t + \phi_1)}\big) \\ &= R_2 i_1 + (l_2 + \tfrac{3}{2}M)p i_r + \tfrac{3}{2}M p\big(\sqrt{2}|\dot{I}_1| \, \mathrm{e}^{\mathrm{j}(s\omega t + \phi_1)}\big). \end{aligned} \tag{4.5}$$

Here $s = (\omega - \omega_m)/\omega$ is the slip. Equation (4.5) contains phases a and r only,

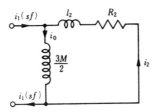

Fig. 4.1 Transient-state equivalent circuit for current-input control of the induction motor

these being segregated from other phases. The phase segregation arises from the spiral vector representation of variables. Equation (4.5) contains the slip frequency but does not explicitly contain the motor speed ω_m. This means that the motor's electromagnetic transient could be independent of the motor speed. This independence is very important in the control theory for motor torque because it means that the electromagnetic transient within the motor can be analysed separately from the dynamic system driven by the motor.

Let the subscript r be changed to 2, in order for phase r to represent the three phases of the secondary. Then equation (4.5) becomes

$$R_2 i_2 + (l_2 + \tfrac{3}{2}M)p i_2 = -\tfrac{3}{2}Mp(\sqrt{2}|\dot{I}_1|\, e^{j(s\omega t + \phi_1')}). \qquad (4.6)$$

This is the performance equation for current-input control of the induction motor. Figure 4.1 shows the corresponding equivalent circuit. The performance equation and its equivalent circuit have been derived without any variable transformation. Their variables are the original variables expressed as spiral vectors. The solution of this performance equation is as simple and easy as its derivation. Its characteristic root is given by

$$p = -\lambda_c = -\frac{R_2}{l_2 + \tfrac{3}{2}M}. \qquad (4.7)$$

The general solution is

$$i_2 = A_2\, e^{-\lambda_c t} + \sqrt{2}|\dot{I}_2|\, e^{j(s\omega t + \phi_2)}. \qquad (4.8)$$

Here A_2 is an arbitrary constant to be determined from an initial condition. The second term is the steady-state current expressed as a circular vector and is given by

$$\left. \begin{aligned} i_2 &= \frac{\tfrac{3}{2}js\omega M \dot{I}_1}{R_2 + js\omega(l_2 + \tfrac{3}{2}M)} = \frac{\tfrac{3}{2}js\omega M}{|Z_2|}|\dot{I}_1|\, e^{j(s\omega t + \phi_1 - \theta_2)}, \\ Z_2 &= R_2 + js\omega(l_2 + \tfrac{3}{2}M) = |Z_2|\, e^{j\theta_2}. \end{aligned} \right\} \qquad (4.9)$$

The right-hand side of equation (4.6) is the input voltage for the equivalent circuit of Fig. 4.1, so input power to the secondary

$$P_2 = \text{Re}\left[-\tfrac{3}{2}Mjs\omega\sqrt{2}I_1\,e^{js\omega t}\right]\text{Re}[i_2] = \tfrac{3}{2}Ms\omega\,\text{Re}[-ji_1]\,\text{Re}[i_2]. \quad (4.10)$$

The revolving speed of the primary magnetomotive force relative to the rotor is $2s\omega/P$ [rad s^{-1}] and the torque per phase is

$$t_1 = \frac{P}{2s\omega}P_2 = \tfrac{3}{4}MP\,\text{Re}[-ji_1]\,\text{Re}[i_2] \qquad [\text{N m}]. \qquad (4.11)$$

Here, P is the number of poles. The three-phase torque t_3 is not $3t_1$ but

$$t_3 = \tfrac{9}{8}MP\,\text{Im}[i_1 i_2^*] \qquad [\text{N m}], \qquad (4.12)$$

where $\text{Im}[z]$ denotes the (real) coefficient of the imaginary part of z and $*$ indicates the complex conjugate. This equation is derived in Appendix III. The three-phase output power is then

$$p_3 = \omega_m t_3, \qquad (4.13)$$

where ω_m is the angular speed in radians/second

When all currents are zero at $t = 0$, current-control input is applied at the primary three-phase terminals. The torque response is then given by

$$t_3 = \tfrac{9}{4}MP|I_1|^2\left(\frac{\tfrac{3}{2}s\omega M}{|Z_2|}\cos\theta_2 - \frac{\tfrac{3}{2}M}{l_2 + \tfrac{3}{2}M}\frac{R_2}{|Z_2|}e^{-\lambda_c t}\sin(s\omega t + \theta_2)\right). \quad (4.14)$$

The derivation of this equation is given in Appendix III.

It should be noted here that this transient torque solution contains the slip frequency but does not contain the motor speed ω_m. This means that transient torque could be independent of motor speed.

When the induction motor in Table 3.1 had no current flow, a primary current $|I_1| = 28$ A was suddenly applied. Currents and torque responses were calculated by equations (4.8) and (4.14) and these are shown in Fig. 4.2. The torque response has large lag and overshoot, that is, poor torque response. This transient will now be eliminated to obtain a very fast torque response.

The exciting current i_0 is given by

$$i_0 = i_1 + i_2 = A_2\,e^{-\lambda t} + \sqrt{2}|I_0|\,e^{j(s\omega t + \phi_0)} \qquad (4.15)$$

The second term is the steady-state exciting current. At $t = 0$, equation (4.15) becomes

$$i_0 = A_2 + \sqrt{2}|I_0|\,e^{j\phi_0}. \qquad (4.16)$$

If

$$i_0 = \sqrt{2}|I_0|\,e^{j\varphi_0} \quad \text{at } t = 0, \qquad (4.17)$$

then A_2 becomes zero. Then the transient does not occur and the torque response becomes instantaneous. This means that if the motor is pre-excited

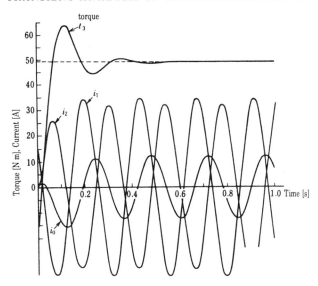

Fig. 4.2 Response of currents and torque for current-input control of the induction motor

with a steady-state exciting current, the transient disappears. This is the principle of pre-excitation, which is the fundamental principle of FAM control. There is one problem in performing the pre-excitation of FAM control. That is the discontinuity of currents in current-input control. When the primary current i_1 is suddenly applied in Fig. 4.1, $i_0(0-)$ is not equal to $i_0(0+)$. In equation (4.16) i_0 is $i_0(0+)$, which must be equal to $\sqrt{2}|I_0|\,e^{j\phi_0}$. This problem is eliminated in T-I type FAM control, as demonstrated in Section 6.1.

4.2.2 *Transient analysis of voltage-input control of the induction motor*

In solving a differential equation, a type of solution is assumed at the start, and if a derived solution is of the assumed type this is the desired solution. An exponential function is often assumed with this method. Thus the spiral vector is the most suitable type of solution for this method. The general solution of equation (4.8) illustrates this fact.

The induction motor model in Fig. 3.1 is symmetrical in structure, so the three phases are expected to have common characteristic roots. Under symmetrical operation, variables of the three phases are also symmetrical and their spiral vector solutions can be assumed as follows:

$$i_a = A_1\,e^{\delta t}, \qquad i_b = A_1\,e^{\delta t - \frac{2}{3}j\pi}, \qquad i_c = A_1\,e^{\delta t + \frac{2}{3}j\pi}. \qquad (4.18)$$

The characteristic root here is

$$\delta = -\lambda + j\omega. \tag{4.19}$$

The characteristic roots of the secondary currents are generally different and are denoted δ'. They are represented by the following spiral vectors:

$$i_r = A_2 e^{\delta' t}, \qquad i_s = A_2 e^{\delta' t - \frac{2}{3}j\pi}, \qquad i_t = A_2 e^{\delta' t + \frac{2}{3}j\pi}. \tag{4.20}$$

The primary currents in equations (4.18) and the secondary currents in equations (4.20) are different from the currents in equations (3.10) and (3.11) only in respect of δ and δ'. And it can then be proved in the same way that equations (3.12)–(3.24) also hold for the currents in equations (4.18) and (4.20). Since these equations are important parts of the analysis, their derivation will be reproduced below.

In this case also, the starting equations are equations (3.6) and (3.7), which are valid for both the steady and transient states. Flux linkages λ_{ga} and λ_{gr} are also given by equations (3.8) and (3.9). For currents in equations (4.18) and (4.20), the following equations hold:

$$i_b - i_c = -j\sqrt{3}i_a, \tag{4.21}$$

$$i_s - i_t = -j\sqrt{3}i_r. \tag{4.22}$$

These are the same as equations (3.13) and (3.14). Inserting these equations in equations (3.8) and (3.9), we get

$$\lambda_{ga} = \tfrac{3}{2}Mi_a + \tfrac{3}{2}Mi_r e^{j\theta}, \tag{4.23}$$

$$\lambda_{gr} = \tfrac{3}{2}Mi_r + \tfrac{3}{2}Mi_a e^{-j\theta}. \tag{4.24}$$

Inserting these equations in equations (3.6) and (3.7), we have

$$v_a = R_1 i_a + l_1 p i_a + \tfrac{3}{2}M[p i_a + p(i_r e^{j\theta})], \tag{4.25}$$

$$0 = R_2 i_r + l_2 p i_r + \tfrac{3}{2}M[p i_r + p(i_a) e^{-j\theta} - j\omega_m i_a e^{-j\theta}], \tag{4.26}$$

where ω_m is angular velocity of the motor given by

$$\omega_m = \frac{d\theta}{dt} = p\theta \qquad [\text{erad s}^{-1}]. \tag{4.27}$$

Multiplying equation (4.26) by $e^{j\theta}$, we get

$$0 = R_2 i_r e^{j\theta} + (l_2 + \tfrac{3}{2}M)(p i_r) e^{j\theta} + \tfrac{3}{2}M(p - j\omega_m)i_a \tag{4.28}$$

Making the variable replacement

$$i_r' = i_r e^{j\theta} = i_r e^{j\omega_m t}, \tag{4.29}$$

we get

$$p i_r' = (p i_r) e^{j\theta} + j\omega_m i_r'. \tag{4.30}$$

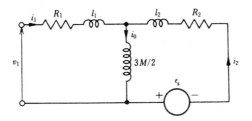

Fig. 4.3   T type transient-state equivalent circuit of the induction motor

Inserting equations (4.29) and (4.30) in equations (4.25) and (4.26), we get

$$v_a = R_1 i_a + (l_1 + \tfrac{3}{2}M)p i_a + \tfrac{3}{2}M p i'_r, \tag{4.31}$$

$$0 = R_2 i'_r + (l_2 + \tfrac{3}{2}M)(p - j\omega_m)i'_r + \tfrac{3}{2}M(p - j\omega_m)i_a. \tag{4.32}$$

These equations are the same as equations (3.23) and (3.24) respectively, where the variables are segregated. By changing subscript a to 1 and i'_r to i_2, equations (4.31) and (4.32) are given by the following matrix equation:

$$\begin{bmatrix} v_1 \\ 0 \end{bmatrix} = \begin{bmatrix} R_1 + (l_1 + \tfrac{3}{2}M)p & \tfrac{3}{2}Mp \\ \tfrac{3}{2}M(p - j\omega_m) & R_2 + (l_2 + \tfrac{3}{2}M)(p - j\omega_m) \end{bmatrix} \begin{bmatrix} i_1 \\ i_2 \end{bmatrix}. \tag{4.33}$$

Equation (4.33) is a very important performance equation for the induction motor and is valid for both transient and steady states. Its variables are the original variables of the motor, and they must be expressed as spiral vectors. As will be shown, solutions of equation (4.33) are spiral vectors, as was assumed in equations (4.18) and (4.20).

Figure 4.3 shows the equivalent circuit corresponding to equation (4.33). In the figure, i_0 is the exciting current given by

$$i_0 = i_1 + i_2. \tag{4.34}$$

The exciting current is identified only through the equivalent circuit. The speed voltage e_s in the equivalent circuit is given by

$$e_s = -j\omega_m[\tfrac{3}{2}M i_1 + (l_2 + \tfrac{3}{2}M)i_2], \tag{4.35}$$

which is proportional to motor angular speed ω_m.

This equivalent circuit is called the T type transient-state equivalent circuit, and it is also valid for the steady state. In the steady state, spiral vectors become circular vectors, for which p becomes $j\omega$. Then matrix equation (4.33) becomes equations (3.30) and (3.31) and the equivalent circuit in Fig. 4.3 becomes the T type steady-state equivalent circuit shown in Fig. 3.2.

The instantaneous values of output power p_0 and torque t_1 per phase are

$$p_1 = \text{Re}[e_s]\,\text{Re}[i_2], \tag{4.36}$$

$$t_1 = \frac{Pp_1}{2\omega_m} \quad [\text{N m}]. \tag{4.37}$$

The three-phase values are not three times p_1 and t_1, but are given by

$$p_3 = \sum_{a,b,c} p_1 = p_{1a} + p_{1b} + p_{1c}, \tag{4.38}$$

$$t_3 = \sum_{a,b,c} t_1 = t_{1a} + t_{1b} + t_{1c} \quad [\text{N m}]. \tag{4.39}$$

This motor torque t_3 is given by

$$t_3 = \tfrac{9}{8}MP\,\text{Im}[i_1 i_2^*] \quad [\text{N m}], \tag{4.40}$$

which is the same as equation (4.12). The three-phase output power is

$$p_3 = \frac{2\omega_m}{P} t_3 = \tfrac{9}{4}M\omega_m\,\text{Im}[i_1 i_2^*] \tag{4.41}$$

The general solution of equation (4.33) will be now sought. Its characteristic equation is

$$\begin{vmatrix} R_1 + (l_2 + \tfrac{3}{2}M)p & \tfrac{3}{2}Mp \\ \tfrac{3}{2}M(p - j\omega_m) & R_2 + (l_2 + \tfrac{3}{2}M)(p - j\omega_m) \end{vmatrix} = 0. \tag{4.42}$$

This is of second degree with respect to p and can be easily solved for p. Let two characteristic roots be denoted by δ_1 and δ_2. Then the general solution of equation (4.33) is given by

$$i_1 = A_1 e^{\delta_1 t} + A_2 e^{\delta_2 t} + \sqrt{2}|\dot{I}_1| e^{j(\omega t + \phi_1)}. \tag{4.43}$$

The first two terms are transient terms, and A_1 and A_2 are arbitrary constants, which are to be determined by initial conditions. They are spiral vectors, as was assumed in equations (4.18) and (4.20). The third term is the steady-state solution for

$$v_1 = \sqrt{2}|\dot{V}_1| e^{j(\omega t - \phi)}. \tag{4.44}$$

All terms of the solution in equation (4.43) are spiral vectors, and the steady-state term is also a circular vector.

The secondary current is obtained by inserting equation (4.43) into the second line of equation (4.33), as follows:

$$i_2 = -\frac{\tfrac{3}{2}M(\delta_1 - j\omega_m)}{R_2 + (l_2 + \tfrac{3}{2}M)(\delta_1 - j\omega_m)} A_1 e^{\delta_1 t} - \frac{\tfrac{3}{2}M(\delta_2 - j\omega_m)}{R_2 + (l_2 + \tfrac{3}{2}M)(\delta_2 - j\omega_m)} A_2 e^{\delta_2 t}. \tag{4.45}$$

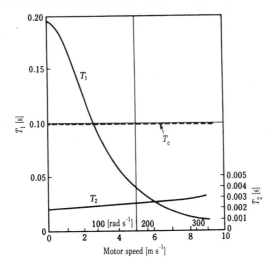

Fig. 4.4 Decay time constants ($T_c = 1/\lambda_c$ in equation (4.7))

For the induction motor in Table 3.1, the transient phenomena were calculated by the solutions in equations (4.43) and (4.45). We will first investigate the characteristic roots. These are expressed as

$$\delta_i = -\frac{1}{T_i} + j\omega_i \qquad (i = 1, 2). \tag{4.46}$$

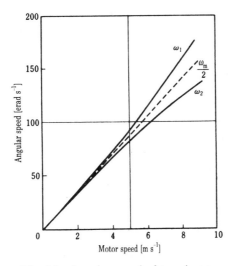

Fig. 4.5 Angular speed of transient terms

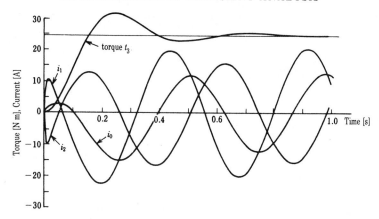

Fig. 4.6 Transient response of the induction motor at stand still, without current at $t = 0$

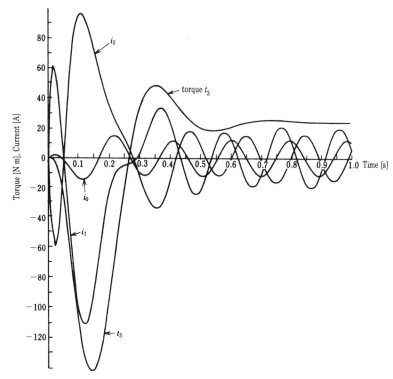

Fig. 4.7 Transient response of the induction motor at no-load speed of 50 Hz, without current at $t = 0$

Here, T_1 and T_2 are decaying time constants, ω_1 and ω_2 are angular velocities in electrical radians/second, and the following relation holds for them, as can be proved by equation (4.42):

$$\omega_1 + \omega_2 = \omega_m. \tag{4.47}$$

Figures 4.4 and 4.5 show the result of calculating the characteristic roots as functions of the motor speed. Time constant T_1 is large and changes considerably, while T_2 is very small and does not change much. In the low-speed range $\omega_1 = \omega_2 = \frac{1}{2}\omega_m$, and in the high-speed range $\omega_1 \rightarrow \omega_m$ and $\omega_2 \rightarrow 0$. In Fig. 4.4, the decaying time constant $T_c = 1/\lambda_c$ of the current-input control in equation (4.7) is also shown. Transient terms with large time constants should be eliminated in order to obtain a fast control response.

When the motor in Table 3.1 at standstill without currents was suddenly impressed with 10.9 V at 2.09 Hz, which correspond to the rated torque, transient current and torque were calculated using equations (4.12), (4.43), and (4.45). The result is shown in Fig. 4.6. Transient currents depend on the phase angle of $\dot{V}_1$ at $t = 0$, but the transient torque does not. The torque transient, though, is large and therefore unsuitable for control purposes. When the motor was running under no load at 50 Hz, the voltage for the rated torque (200 V, 52.09 Hz) was suddenly impressed. The transient was calculated and the result is shown in Fig. 4.7. A very severe transient occurs, and peak values of currents and torque are very large, 5–6 times their rated values.

As shown above, the transient phenomena for the initial condition of zero currents are very severe, and they are not suitable for control purposes. The remedy for this is FAM control, which will be explained in Chapter 6.

5 Transformation of transient-state equivalent circuits of the induction motor

In Section 3.5, the steady-state equivalent circuit of the induction motor was transformed to derive T-I and T-II type equivalent circuits from T type equivalent circuit. Although they have the same input-terminal impedance, FAM control derived different features from the different equivalent circuits. The same variable transformation will now be applied to T type transient-state equivalent circuit to derive T-I and T-II type transient-state equivalent circuits.

Currents are now transformed by the following matrix equation:

$$\begin{bmatrix} i_1 \\ i_2 \end{bmatrix} = \begin{bmatrix} 1 & 0 \\ 0 & \alpha \end{bmatrix}\begin{bmatrix} i_1 \\ i_2^\alpha \end{bmatrix}. \tag{5.1}$$

This is the same transformation as equation (3.49). Equation (4.33) is now transformed according to equation (3.50), to get

$$\begin{bmatrix} v_1 \\ 0 \end{bmatrix} = \begin{bmatrix} R_1 + (l_1 + \tfrac{3}{2}M)p & \tfrac{3}{2}M\alpha p \\ \tfrac{3}{2}M\alpha(p - j\omega_m) & R_2\alpha^2 + \alpha^2(l_2 + \tfrac{3}{2}M)(p - j\omega_m) \end{bmatrix}\begin{bmatrix} i_1 \\ i_2^\alpha \end{bmatrix}. \tag{5.2}$$

Figure 5.1 shows the corresponding equivalent circuit, where the speed voltage e_s^α is given by

$$e_s^\alpha = -\tfrac{3}{2}j\omega_m M\alpha i_1 - j\omega_m \alpha^2(l_2 + \tfrac{3}{2}M)i_2^\alpha \tag{5.3}$$

This is the general transient-state equivalent circuit, containing the arbitrary constant α. There are therefore an infinite number of equivalent circuits that have the same transient input impedance at the primary terminals. This can be proved in the same way as equation (3.52). Electromagnetic transient phenomena within the motor, however, differ for different values of α, giving different equivalent circuits. We shall now derive suitable equivalent circuits by choosing appropriate values for α.

When α is chosen as

$$\alpha = \frac{\tfrac{3}{2}M}{l_2 + \tfrac{3}{2}M}, \tag{5.4}$$

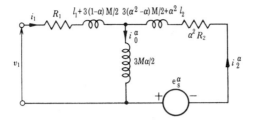

Fig. 5.1   General transient-state equivalent circuit of the induction motor

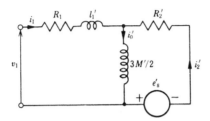

Fig. 5.2   T-I type transient-state equivalent circuit of the induction motor

equation (5.2) becomes

$$
\begin{bmatrix} v_1 \\ 0 \end{bmatrix} = \begin{bmatrix} R_1 + (l_1' + \tfrac{3}{2}M')p & \tfrac{3}{2}M'p \\ \tfrac{3}{2}M'(p - j\omega_m) & R_2' + \tfrac{3}{2}M'(p - j\omega_m) \end{bmatrix} \begin{bmatrix} i_1 \\ i_2' \end{bmatrix}, \tag{5.5}
$$

where $\tfrac{3}{2}M' = \tfrac{3}{2}M\alpha$, $R_2' = \alpha^2 R_2$, and $l_1' = l_1 + \tfrac{3}{2}M(1 - \alpha)$. The corresponding equivalent circuit is given in Fig. 5.2, where the speed voltage e_s' is given by

$$
e_s' = -\tfrac{3}{2}j\omega_m M'(i_1 + i_2') = -\tfrac{3}{2}j\omega_m M' i_0'. \tag{5.6}
$$

It should be noted here that the secondary leakage inductance is zero. This circuit is called the T-I type transient-state equivalent circuit.

When α is chosen as

$$
\alpha = \frac{l_1 + \tfrac{3}{2}M}{\tfrac{3}{2}M}, \tag{5.7}
$$

equation (5.2) becomes equation (5.8).

$$
\begin{bmatrix} v_1 \\ 0 \end{bmatrix} = \begin{bmatrix} R_1 + \tfrac{3}{2}M''p & \tfrac{3}{2}M''p \\ \tfrac{3}{2}M''(p - j\omega_m) & R_2'' + (l_2'' + \tfrac{3}{2}M'')(p - j\omega_m) \end{bmatrix} \begin{bmatrix} i_1 \\ i_2'' \end{bmatrix}. \tag{5.8}
$$

Here, $\tfrac{3}{2}M'' = \tfrac{3}{2}M\alpha$, $R_2'' = \alpha^2 R_2$, and $l_2'' = \alpha^2 l_2 + \tfrac{3}{2}M(\alpha^2 - \alpha)$. Figure 5.3 shows the corresponding equivalent circuit, where the speed voltage e_s'' is given by

$$
e_s'' = -\tfrac{3}{2}j\omega_m M'' i_1'' - j\omega_m(l_2'' + \tfrac{3}{2}M'')i_2''. \tag{5.9}
$$

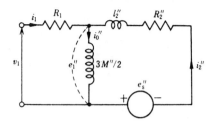

Fig. 5.3 T-II type transient-state equivalent circuit of the induction motor

It should be noted here that the primary leakage inductance l_1'' is zero. This circuit is called the T-II type transient-state equivalent circuit.

In the T-I type transient equivalent circuit of Fig. 5.2, the following approximations hold:

$$l_1' = l_1 + l_2, \qquad \tfrac{3}{2}M' = \tfrac{3}{2}M - l_2. \qquad (5.10)$$

In the T-II type transient equivalent of Fig. 5.3, the following approximations hold:

$$l_2'' = l_1 + l_2, \qquad \tfrac{3}{2}M'' = \tfrac{3}{2}M + l_1. \qquad (5.11)$$

These approximations reveal the nature of the inductances, but they should not be used in control computations. Accurate values of circuit constants should be used instead. For the induction motor in Table 3.1, the values of α for T-I and T-II type circuits are respectively 0.980 and 1.031.

The T type, T-I type, and T-II type equivalent circuits derived above are also valid for steady states. And when $p = j\omega$ is inserted in equations (5.5) and (5.8), the corresponding steady-state equivalent circuits are obtained, as in Chapter 3.

6 Transient analysis and torque control of the induction motor

In the preceding chapter, various transient-state equivalent circuits were derived. In this chapter, they will be used to analyse the transient response of induction motor torque control, and the results obtained will be put to use to provide a very fast torque response. As was shown in Chapter 3, a constant-magnitude exciting current makes the torque–speed characteristic curves very straight over very wide torque ranges. A constant-magnitude exciting current is the fundamental principle behind FAM control of induction motor torque. This principle, coupled with the pre-excitation, suppresses transients and makes control response very fast.

The control input to the induction motor is either primary current or primary voltage. These are AC quantities, which have amplitude, frequency, and phase. Current-input control and voltage-input control are possible, and these are different in terms of input and control methods and responses. The accompanying transients, quickness of response, and hardware used for control are different, and they are fed from three-phase inverters of different types. In the two controls, the power supplies have different internal impedances. In voltage-input control, the internal impedance of the power supply is zero, while it is infinite in current-input control. This difference gives rise to different electromagnetic transients and different characteristics for each of the controls. In addition, the appropriate equivalent circuits on which the controls are based are different.

6.1 Current-input control of induction motor torque

The T-I type transient-state equivalent circuit shown in Fig. 5.2 is very suitable for current-input control of induction motor torque. Its performance equation is given by the second line of equation (5.5), that is,

$$-\tfrac{3}{2}M'(p - \mathrm{j}\omega_{\mathrm{m}})i_1 = R_2'i_2' + \tfrac{3}{2}M'(p - \mathrm{j}\omega_{\mathrm{m}})i_2'. \tag{6.1}$$

This is different in two ways from equation (4.6), which is also a performance equation for current-input control. First here l_2' is zero, because this is T-I type equivalent circuit. Second, the state variables here are observed on the stator side and $i_2' = i_2\,\mathrm{e}^{\mathrm{j}\omega_{\mathrm{m}}t}$ (see equation (3.21)).

In current-input control, i_1 is the control input and therefore the characteristic equation (6.1) is

$$R_2' + \tfrac{3}{2}M'(p - j\omega_m) = 0,\qquad(6.2)$$

and its characteristic root is

$$p = -\frac{R_2'}{\tfrac{3}{2}M'} + j\omega_m = -\frac{\alpha^2 R_2}{\tfrac{3}{2}M\alpha} + j\omega_m = -\frac{R_2}{l_2 + \tfrac{3}{2}M} + j\omega_m = -\lambda_c + j\omega_m,\qquad(6.3)$$

where α is given by equation (5.4) and λ_c by equation (4.7). This means that with respect to the characteristic root the two equivalent circuits of T and T-I types are equivalent. The general solution of equation (6.1) is

$$i_2' = A_2\,e^{-\lambda_c t}\,e^{j\omega_m t} + \sqrt{2}|\dot{I}_2'|\,e^{j(\omega t + \phi_2)},\qquad(6.4)$$

where A_2 is an arbitrary constant which is determined from an initial condition, and the second term is a steady-state term expressed as a circular vector. For the motor in Table 3.1, the decay time constant is $T_c = 1/\lambda_c = 0.1$ s, which is rather large. It is therefore necessary to make A_2 as small as possible in order to obtain a fast response.

The exciting current i_0' in the equivalent circuit of Fig. 5.2 is

$$i_0' = i_1 + i_2' = \sqrt{2}|\dot{I}_1|\,e^{j(\omega t + \phi_1)} + \sqrt{2}|\dot{I}_2'|\,e^{j(\omega t + \phi_2)} + A_2\,e^{-\lambda_c t}\,e^{j\omega_m t}$$

$$= \sqrt{2}|\dot{I}_0'|\,e^{j(\omega t + \phi_0)} + A_2\,e^{-\lambda_c t}\,e^{j\omega_m t}.\qquad(6.5)$$

The first term is the steady-state exciting current. At $t = 0$, equation (6.5) becomes

$$i_0'(t = 0) = \sqrt{2}|\dot{I}_0'|\,e^{j\phi_0} + A_2.\qquad(6.6)$$

When the equation

$$i_0'(t = 0) = \sqrt{2}|\dot{I}_0'|\,e^{j\phi_0}\qquad(6.7)$$

holds at $t = 0$, then equation (6.6) gives $A_2 = 0$, and the transient does not appear in the control. This means that if the motor is pre-excited before $t = 0$ with a steady-state exciting current, and the primary current i_1, as control input, contains this value, then the transient does not occur. This is the principle of pre-excitation, which was briefly mentioned in Section 4.2.1.

When the torque is zero, only the exciting current flows during pre-excitation. There are two ways for the exciting current to flow. Since $i_0' = i_1 + i_2'$, it follows that i_0' flows as primary current or secondary current. When the torque and secondary current are zero, the exciting current flows as primary current. When the primary current is suddenly switched off, the exciting current flows as secondary current. There is also the case when the exciting current flows partly as primary current and partly as secondary current.

The Laplace transformation may be used here to further clarify the situation. Inserting $i'_0 = i_1 + i'_2$ in equation (6.1), we get

$$R'_2 i'_1 = R'_2 i'_0 + \tfrac{3}{2}M'(p - j\omega_m)i'_0. \tag{6.8}$$

Performing the Laplace transformation, we obtain

$$R'_2 I_1(s) = R'_2 I'_0(s) + \tfrac{3}{2}M'[sI'_0(s) - I'_{00}] - \tfrac{3}{2}j\omega_m M'I'_0(s). \tag{6.9}$$

Here, s is a complex variable of the transformation and I'_{00} is the initial value of i'_0 at $t = 0$. Rewriting this, we get

$$I'_0(s) = \frac{R'_2}{\tfrac{3}{2}M'} \frac{I_1(s)}{s + R'_2/\tfrac{3}{2}M' - j\omega_m} + \frac{I'_{00}}{s + R_2/\tfrac{3}{2}M' - j\omega_m}. \tag{6.10}$$

The characteristic root of this equation is the same as that given by equation (6.3). The primary current i_1 is the control input and is given by the following circular vector:

$$i_1 = \sqrt{2|I_1|}\, e^{j(\omega t + \phi_1)}. \tag{6.11}$$

Its Laplace transformation is

$$I_1(s) = \frac{\sqrt{2|I_1|}\, e^{j\phi_1}}{s - j\omega}. \tag{6.12}$$

Inserting this into equation (6.9), we have

$$I'_0(s) = \frac{R'_2}{\tfrac{3}{2}M'} \frac{\sqrt{2|I_1|}\, e^{j\phi_1}}{(s - \delta)(s - j\omega)} + \frac{I'_{00}}{s - \delta}. \tag{6.13}$$

Here δ is the characteristic root p of equation (6.3). Performing the inverse Laplace transformation, we obtain

$$i'_0 = R'_2 \frac{\sqrt{2|I_1|}\, e^{j(\omega t + \phi_1)}}{R'_2 + \tfrac{3}{2}js\omega M'} - R'_2 \frac{\sqrt{2|I_1|}\, e^{j\phi_1}}{R'_2 + \tfrac{3}{2}js\omega M'} e^{\delta t} + I'_{00}\, e^{\delta t}. \tag{6.14}$$

The first term is a steady-state exciting current. Let this be denoted by

$$R'_2 \frac{\sqrt{2|I_1|}\, e^{j(\omega t + \phi_1)}}{R'_2 + \tfrac{3}{2}js\omega M'} = \sqrt{2|I'_0|}\, e^{j(\omega t + \phi_0)}. \tag{6.15}$$

Then equation (6.14) can be rewritten as

$$i'_0 = \sqrt{2|I'_0|}\, e^{j(\omega t + \phi_0)} - \sqrt{2|I'_0|}\, e^{j\phi_0} e^{\delta t} + I'_{00}\, e^{\delta t}. \tag{6.16}$$

When the equation

$$\sqrt{2|I'_0|}\, e^{j\phi_0} = I'_{00}. \tag{6.17}$$

holds, the second and third terms of equation (6.14), which are transient terms, cancel each other out. This means that when the initial value I'_{00} of the exciting current is equal to the steady-state exciting current at $t = 0$, no

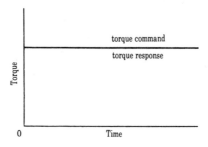

Fig. 6.1 Response of T-I type FAM control (current-input control) of induction motor torque

transient occurs. If the exciting current is thereafter kept as a constant-amplitude circular vector, the transient phenomenon does not occur. One advantage of the T-I equivalent circuit is that $i'_0(0-) = i'_0(0+)$ holds because of the zero secondary leakage inductance. As explained in Section 4.2.1, in T type equivalent circuit $i'_0(0-) \neq i'_0(0+)$. Applying the Laplace transformation to equation (4.6), we can check this point and obtain equation (4.14).

The current-input control of induction motor torque is performed by the set-up shown in Fig. 3.19. In the figure, the exciting current $|I'_0|$, which is fixed at a rated value and the secondary current $|I'_2|$, which is proportional to the torque command T_c, are fed to the vector adder, to produce the primary current $|I_1| e^{j\phi_1}$. Then $|I_1| e^{j\phi_1}$ and f are fed to the unit vector generator to obtain the instantaneous value of primary current i_1, which is fed to drive the three-phase inverter. This control satisfies the pre-excitation condition and keeps the exciting current $|I'_0|$ constant, so a transient does not occur, and the torque response becomes instantaneous, as shown in Fig. 6.1.

There are some problems with current-input control. One is that the primary current must rise vertically, as shown in Fig. 6.2. Consequently,

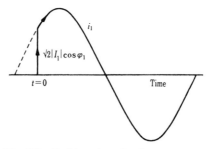

Fig. 6.2 Sudden rise of primary current

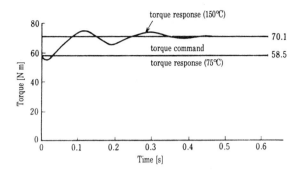

Fig. 6.3   Torque response of current-input control of induction motor torque

di_1/dt becomes infinite and involves an infinite voltage of $l_1'\, di_1/dt$ in the equivalent circuit of Fig. 5.2. However, the primary leakage inductance l_1' is so small that i_1 can rise to $\sqrt{2}|I_1|\cos\phi_1$ at $t = 0$ within several milliseconds without causing a very high voltage $l_1\, di_1/dt$.

When there is some error in circuit constant values, the condition of equation (6.17) for no transient is not satisfied and transients occur. Secondary resistance change due to rotor temperature variation causes such errors. Figure 6.3 shows the influence of secondary resistance change. When the secondary conductor temperature is 75°C, R_2' has the right value and torque control response is instantaneous. When it becomes 150°C, R_2' increases by about 30%, the torque response has a transient and the steady-state torque increases by about 20% of the rated torque for the motor in Table 3.1. The influence of secondary resistance change must be compensated for. This will be explained in Section 6.5.

6.2   Voltage-input control of induction motor torque

Based on the T-II transient-state equivalent circuit in Fig. 5.3, voltage-input control of induction motor torque will now be explained. The following equations hold:

$$e_1'' = \tfrac{3}{2}M''pi_2'', \tag{6.18}$$

$$e_1'' = -R_2''i_2'' - l_2''(p - j\omega_m)i_2'' + \tfrac{3}{2}j\omega_m M''i_0'', \tag{6.19}$$

$$v_1 = R_1i_1 + e_1''. \tag{6.20}$$

Since in FAM control the exciting current i_0'' is a circular vector, e_1'' in equation (6.19) becomes the following circular vector:

$$e_1'' = \tfrac{3}{2}j\omega M''i_0'' = \tfrac{3}{2}j\omega M''\sqrt{2}|\dot{I}_0''|\, e^{j(\omega t + \phi_0)}$$

$$= \sqrt{2}|\dot{E}_1''|\, e^{j(\omega t + \phi_0 + \pi/2)}, \qquad |E_1''| = \tfrac{3}{2}\omega\sqrt{2}|\dot{I}_0''|M''. \tag{6.21}$$

Here, i_0'' is the following circular vector:

$$i_0'' = \sqrt{2}|i_0''| \, e^{j(\omega t + \phi_0)}. \tag{6.22}$$

Inserting equation (6.22) into equation (6.19), we get

$$e_1''(1 - \omega_m/\omega) = se_1'' = -[R_2'' + l_2''(p - j\omega_m)]i_2''. \tag{6.23}$$

For the steady-state solution of equation (6.23), p is set to $j\omega$ and we get

$$i_{2s}'' = -\frac{\sqrt{2}|E_1''|}{R_2''/s + j\omega l_2} \, e^{j(\omega t + \phi_0 + \frac{1}{2}\pi)}. \tag{6.24}$$

The general transient solution of equation (6.23) is

$$i_{2t}'' = A \, e^{-(R_2''/l_2'')t} \, e^{j\omega_m t}. \tag{6.25}$$

The general solution is

$$i_2'' = i_{2s}'' + i_{2t}'' = -\frac{\sqrt{2}|E_1''|}{R_2''/s + j\omega l_2} \, e^{j(\omega t + \phi_0 + \frac{1}{2}\pi)} + A \, e^{-(R_2''/l_2'')t} \, e^{j\omega_m t}$$

$$= \sqrt{2}|I_2''| \, e^{j(\omega t + \phi_2)} + A \, e^{-(R_2''/l_2'')t} \, e^{j\omega_m t} \tag{6.26}$$

The decay time constant is

$$T_v'' = \frac{l_2''}{R_2''} = \frac{l_2 + l_1}{R_2}, \tag{6.27}$$

which is very small. For the motor in Table 3.1, $T_v'' = 0.0049$ s, while the time constant $T_c = 1/\lambda_c$ for the current-input control of equation (4.7) is 0.1 s.

When the motor is running under no load with rated voltage and rated frequency, the torque command of the rated value is suddenly given. Subsequently, frequency is increased by 2.5 Hz. For the initial condition of $i_2'' = 0$, equation (6.26) gives

$$A = -\sqrt{2}|I_2''| \, e^{j\phi_2}. \tag{6.28}$$

Inserting this into equation (6.26), we get

$$i_2'' = \sqrt{2}|I_2''| \, e^{j(\omega t + \phi_2)} - \sqrt{2}|I_2''| \, e^{-(R_2''/l_2'')t} \, e^{j\omega_m t}. \tag{6.29}$$

The three-phase torque is given by

$$t_3 = \tfrac{9}{8}MP \, \mathrm{Im}[i_0'' i_2''^*]$$

$$= \tfrac{9}{8}MP \, \mathrm{Im}[\sqrt{2}|I_0''| \, e^{j(\omega t + \phi_0)}$$

$$\times (\sqrt{2}|I_2''| \, e^{-j(\omega t + \phi_2)} - \sqrt{2}|I_2''| \, e^{-(R_2''/l_2'')t} \, e^{-j(\omega_m t + \phi_2)})]$$

$$= \tfrac{9}{4}MP|I_0''| \, |I_2''|[\sin(\phi_0 - \phi_2) - e^{-(R_2''/l_2'')t} \sin(s\omega t + \phi_0 - \phi_2)]. \tag{6.30}$$

From equation (6.24), the following relation holds

$$\phi_2 = \phi_0 - \theta_2 + \tfrac{1}{2}\pi, \qquad \theta_2 = \tan^{-1}(s\omega l_2''/R_2''). \tag{6.31}$$

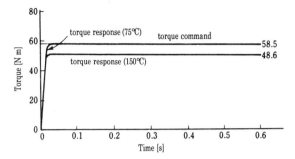

Fig. 6.4 Response of voltage-input FAM control of induction motor torque

Therefore, equation (6.30) becomes

$$t_3 = \tfrac{9}{4}MP|\dot{I}_0''|\,|\dot{I}_2''|[\cos\theta_2 - e^{-(R_2''/l_2'')t}\cos(s\omega t + \theta_2)].\tag{6.32}$$

Figure 6.4 shows the torque response of equations (6.30) and (6.32). Response is not instantaneous, but it is very fast. The influence of secondary resistance change due to temperature variation is also shown in this figure. When the secondary conductor temperature rises to 150°C from 75°C, the steady-state torque decreases by about 20%. But still its response is very fast.

Voltage-input control will be now explained based on the T type transient-state equivalent circuit shown in Fig. 4.3. In this circuit, the following equations hold:

$$e_1 = \tfrac{3}{2}Mpi_0,\tag{6.33}$$

$$e_1 = -R_2 i_2 - l_2(p - j\omega_m)i_2 + \tfrac{3}{2}j\omega_m M i_0,\tag{6.34}$$

$$v_1 = R_1 i_1 + l_1 p i_1 + e_1.\tag{6.35}$$

In FAM control, the exciting current must be kept as a constant-amplitude circular vector, as follows:

$$i_0 = \sqrt{2}|\dot{I}_0|\,e^{j(\omega t + \phi_0)}.\tag{6.36}$$

Similarly to equation (6.23), we get

$$se_1 = -[R_2 + l_2(p - j\omega_m)]i_2.\tag{6.37}$$

The general solution of this equation is

$$i_2 = \sqrt{2}|\dot{I}_2|\,e^{j(\omega t + \phi_2)} + A\,e^{-(R_2/l_2)t}\,e^{j\omega_m t}.\tag{6.38}$$

Here, the first term is the steady-state secondary current expressed as a circular vector. For the initial condition $i_2 = 0$ at $t = 0$, the three-phase

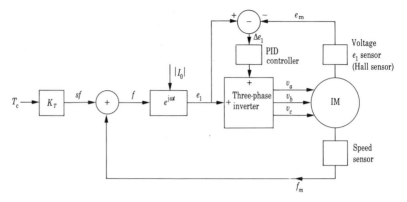

Fig. 6.5 Block diagram of voltage-input T type FAM control of induction motor torque

torque is

$$t_3 = \tfrac{9}{4}MP|\dot{I}_0|\,|\dot{I}_2|[\sin(\phi_0 - \phi_2) - e^{-(R_2/l_2)t}\sin(s\omega t + \phi_0 - \phi_2)], \quad (6.39)$$

which is similar to equation (6.30). This is rewritten as

$$t_3 = \tfrac{9}{4}MP|\dot{I}_0|\,|\dot{I}_2|[\cos\theta_2 - e^{-(R_2/l_2)t}\cos(s\omega t + \theta_2)], \quad (6.40)$$

which is similar to equation (6.32).

The torque response of equations (6.39) and (6.40) is about the same as in Fig. 6.4. Again, it is not instantaneous, but it is very fast. The time constant l_2/R_2 is 2 ms for equation (6.40), compared with 5 ms for equation (6.32).

Figure 3.20 shows a block diagram of voltage-input T-II type FAM control of induction motor torque. Figure 6.5 shows a block diagram for voltage-input T type FAM control. A Hall sensor placed in the air gap measures exciting voltage e_m, which is compared with its signal e_1 to obtain the error signal Δe_1. Next, Δe_1 is fed through the PID controller to the inverter, to control the primary voltage v, so that Δe_1 tends toward zero. Thus i_0 is kept as a given circular vector and the voltage-input T type FAM control explained above is performed.

6.3   Transfer function of FAM control of induction motor torque

As shown in Fig. 6.6, the induction motor, as a control motor, drives a machine or equipment, whose state variable x is controlled to attain x_c, the command value. This x can be speed, position, or any other state variable. It is sensed and is fed back to generate the error signal Δx, which is fed through the compensation circuit to generate the torque command for the

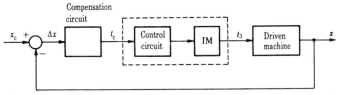

Fig. 6.6 Block diagram of closed-loop control by means of the induction motor

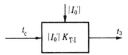

Fig. 6.7 Transfer function for current-input FAM control of induction motor torque

induction motor. This is an outline of closed-loop control. In order to analyse closed-loop control, the induction motor torque control, which is situated in the forward path of the control, must be expressed in terms of its transfer function.

In the current-input T-I type FAM control of induction motor torque, response is instantaneous, as shown in Fig. 6.1. Consequently, the transfer function given in Fig. 6.7 represents the induction motor together with the control circuit surrounded by the broken line in Fig. 6.6. It is a real value proportional to $|I_0'|$, the effective value of the exciting current. This means that field adjustment in the induction motor control is possible, just like with a DC control motor.

In the voltage-input control of T-II type FAM of induction motor torque, response is given by equation (6.32). In this equation, the slip frequency is very low. For example, for the motor in Table 3.1, the slip frequency is 2.4 Hz for the rated torque. The decay time constant $T_v'' = l_2''/R_2''$ is 5 ms, and so, in equation (6.32), $\cos(s\omega t + \theta_2) \approx \cos\theta_2$ for $t < 3T_v''$. Thus equation (6.32) is approximated by the following equation:

$$t_3 = T_c(1 - e^{-(R_2''/l_2'')t}). \qquad (6.41)$$

The response is of first-order lag, as shown in Fig. 6.4, and its transfer function is shown in Fig. 6.8. Here, field adjustment is also possible.

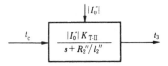

Fig. 6.8 Transfer function for voltage-input FAM control of induction motor torque

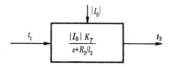

Fig. 6.9   Transfer function for voltage-input T type FAM control of induction motor torque

For the voltage-input T type FAM control of induction motor torque, response is shown in Fig. 6.4. Its transfer function is given in Fig. 6.9; it is the same as that in Fig. 6.8 except that R_2''/l_2'' is replaced by R_2/l_2. This means that T type FAM control responds faster than T-II type FAM control.

6.4   Current- and voltage-input FAM control of induction motor torque

As explained in the preceding sections, current-input control and voltage-input control have different torque control responses, although both are very fast. Their transfer functions are also different. Current-input control can give an instantaneous response, while voltage-input control has a response of first-order lag of very small time constant. In this respect current-input control is the better of the two. However, this is true only when there is no error in control. If there is some error in control computation, the response deteriorates. For example, when there is an error in the secondary resistance value, torque responses deteriorate as shown in Fig. 6.3 for the current-input control and in Fig. 6.4 for the voltage-input control. Although these deteriorations are small, that of the latter is smaller. In this respect, voltage-input control is preferable.

In choosing one of the two controls, the inverter type must be taken into consideration. There are two types of inverter: the voltage-source inverter and the current-source inverter. Both types can feed both types of control. This point will be explained in Chapter 9.

6.5   Temperature compensation for induction motor torque control

FAM control has derived superior control features from the induction motor, as explained in this chapter. It is an open-loop control of induction motor torque, and uses circuit constants in its control computation. Accurate values of the circuit constants are a prerequisite for FAM control. In the equivalent circuits, there are inductances and resistances, which are the circuit constants.

Iron saturation may cause variations in inductance. FAM control, in which the amplitude of the exciting current is kept, however, limits its influence to a minimum. Constant values of inductance give good agreement between measured torque and calculated torque, as shown in Fig. 3.16.

Figures 6.3 and 6.4 show the influence of temperature variation on motor characteristics and control response. A rotor temperature rise from 75°C to 150°C increases the steady-state torque by 20% for current-input control and decreases it by 17% for voltage-input control. And it deteriorates torque response.

6.5.1   Temperature compensation for voltage-input control

The reference temperature for determining motor characterisics is usually 75°C. Secondary resistance at 75°C is now denoted by $R_{2.75}$. As a result, the secondary resistance at t°C is given by

$$R_{2,t} = \frac{1 + t\alpha}{1 + 75\alpha} R_{2,75} = \frac{1/\alpha + t}{1/\alpha + 75} R_{2,75}. \tag{6.42}$$

Here, α is the temperature coefficient of electric resistance. For the aluminium cage rotor, equation (6.42) becomes

$$R_{2,t} = \frac{218 + t}{293} R_{2,75}. \tag{6.43}$$

For the T-I type steady-state equivalent circuit shown in Fig. 3.9, the following equation holds:

$$|\dot{I}_1| = \left(\frac{R_2'^2 + (sx_m')^2}{R_2'^2} \right)^{\frac{1}{2}} |\dot{I}_0'|. \tag{6.44}$$

In FAM control, $|\dot{I}_0'|$ is given and kept constant, so the torque and slip frequency are related by equation (3.59), that is, they are proportional. For the motor in Table 3.1, the primary current and torque were calculated by equations (6.44) and (3.59), respectively, for different temperatures. Results are shown in Figs. 6.10 and 6.11. At 75°C, the primary current and torque are at point A for a slip frequency of 5 Hz. At 150°C they are at point B, where the primary current decreased by 18% and torque decreased by 20%. The error $\Delta|I_1|$ is determined by measuring the primary current of the motor. By increasing the slip frequency in equation (6.44), $\Delta|I_1|$ can be brought to zero. When $\Delta|I_1|$ becomes zero, R_2'/s attains the correct value of R_2'/s. Then $|I_1|$ and the torque are brought back to the level of point A from the level of point B. Thus the correct value of secondary resistance can be determined from the correct value of R_2'/s.

In this way, the influence of rotor temperature variation can be compensated for. Temperature variation is a rather slow phenomenon, and so a control

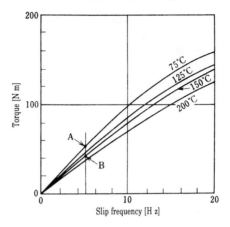

Fig. 6.10 Torque versus slip frequency characteristic curves of the FAM controlled induction motor

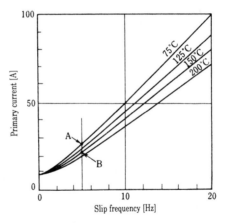

Fig. 6.11 Primary current versus slip frequency characteristic curves of the FAM controlled induction motor

microprocessor can be used for temperature compensation, time-sharing with other control functions.

6.5.2 *Secondary current feedback control of induction motor torque*

It is possible to compensate for rotor temperature change by feeding back the secondary current. A new type of secondary current sensor will be proposed for this purpose. Current-input FAM control will be used as an example.

In the T-I steady-state equivalent circuit of Fig. 3.9, equations (3.58), (3.59), and (3.61) hold. They are rewritten as

$$I_2' = -j \frac{s\omega M'}{R_2'} I_0',$$ (6.45)

$$T = \tfrac{3}{2} P \frac{1}{R_2'} (\tfrac{3}{2} M')^2 s\omega |I_0'|^2,$$ (6.46)

$$T = \tfrac{9}{4} M' P |I_0'| |I_2'|.$$ (6.47)

From these equations, the proportional relation of equation (6.48) below holds for FAM control, where $|I_0'|$ is kept constant. In addition, the torque–speed curves become completely straight, as shown in Fig. 3.16.

$$T \propto sf \propto |I_2'|.$$ (6.48)

When the secondary resistance R_2' changes, the slip frequency sf is adjusted to keep R_2'/sf unchanged, and $|I_2'|$ and T remain unchanged, according to equations (6.45) and (6.47). This is the proportion law for FAM control of induction motor torque. With a constant-voltage constant-frequency drive, the proportion law is that constant R_2/s keeps the torque unchanged. However, in FAM control, the law is changed so that a constant R_2/sf keeps the torque unchanged.

According to equation (6.47), the torque T in FAM control is proportional to $|I_2'|$, irrespective of R_2' and sf, and $|I_2'|$ is proportional to sf. These proportional relations make possible the secondary current feedback compensation and control of the induction motor explained below.

When the motor is the wound rotor type, three-phase secondary currents can be directly measured. Let their instantaneous real values be i_r, i_s, and i_t. Effective values of these currents are given by

$$|I_2| = \frac{1}{\sqrt{2}} [i_r^2 + \tfrac{1}{3}(i_s - i_t)]^{\frac{1}{2}}.$$ (6.49)

When the motor is the cage rotor type, the secondary current i_2 is measured by a stationary sensor placed in the leakage magnetic field produced by the secondary current. The sensor is placed near the revolving short-circuited ring of the rotor, as shown in Fig. 6.12, fixed in space and shielded carefully from the leakage field due to the primary current. When the sensor is a Hall element, it measures flux density b, which is proportional to i_2. Thus its output voltage is

$$e_h = K i_2 = K\sqrt{2}|I_2| e^{j(\omega t + \phi_2)}.$$ (6.50)

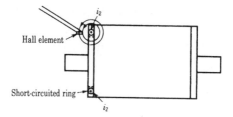

Fig. 6.12   Secondary current sensor placed near the short-circuited ring of the rotor

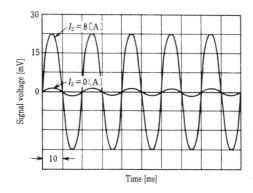

Fig. 6.13   An example of signal voltage of the Hall sensor for secondary current

If a small search coil is used as a sensor, its output voltage is

$$e_c \propto db/dt = K'j\omega\sqrt{2}|I_2|\, e^{j(\omega t + \phi_2)}. \qquad (6.51)$$

Figure 6.13 shows an oscillogram of the Hall sensor, which is a good sensor signal. When two Hall sensors are fixed at quadrature positions in space and their output voltages are e_{h1} and e_{h2}, we get

$$|I_2| \propto (e_{h1}^2 + e_{h2}^2)^{\frac{1}{2}}. \qquad (6.52)$$

As explained in Section 6.1, the response of current-input control of motor torque can be instantaneous, but when the secondary resistance changes because of temperature variation, a transient occurs and the steady-state value of the torque deviates from the torque command. This is now compensated for by secondary current feedback.

Figure 6.14 shows the block diagram for this method. The upper portion is the same as the block diagram of the current-input control shown in Fig. 3.19, to which the secondary current feedback is added in the lower portion. The secondary current signal $|I_2'|_s$ sensed by the secondary current sensor is compared with its signal $|I_2'|$ to obtain the deviation $\Delta|I_2'|$. This is turned

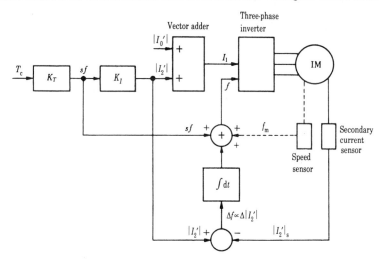

Fig. 6.14   Block diagram of secondary current feedback control of induction motor torque

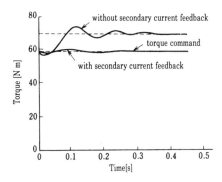

Fig. 6.15   Torque response of secondary current feedback control

into the incremental frequency Δf proportionally; then Δf is integrated in order to be added to the frequency f which is fed to the inverter, This process is continued until $\Delta|I'_2|$ becomes zero, which means that the secondary current becomes equal to torque command. Figure 6.15 shows a computer simulation result of secondary current feedback control when output of the integrator started from zero. The response is fast, and the steady-state torque has no deviation.

In Fig. 6.14 the feedback path of the motor speed is indicated by the broken line. This path, including the speed sensor, can be omitted, because motor speed signal f_m is included and memorized in the output of the integrator of Δf. The secondary current feedback method explained above eliminates the influence of secondary resistance variation and dispenses with the speed sensor, which is an expensive component.

7 Analysis and control of the synchronous motor as a control motor

7.1 The permanent-magnetic-excited synchronous motor

The synchronous motor has been used for many years as a constant speed motor. Owing to remarkable advancements in inverter technique, inverter output voltage, current, and frequency can now be controlled very easily and quickly, and thus the synchronous motor is now used more and more as a fast response servomotor. In fast response applications, torque is the control output of the motor, as it was for the induction motor in the preceding chapters. As fast response motors, synchronous motors are now replacing DC motors, driving robots and numerically controlled machine tools, etc. These synchronous motors are rather small, usually less than several tens of kilowatts. However, the analytical theory is common to both small and large motors. The following analysis and control theory are applicable, irrespective of motor size.

Quick response control synchronous motors or synchronous servomotors are mostly of the permanent-magnet-excited type. Figure 7.1 depicts the rotors of these motors. In Fig. 7.1(a, b) permanent magnets are placed on the rotor surface, and in Fig. 7.1(c) they are placed at inner positions of the rotor core. The surface-magnet type increases effective gap length, resulting in smaller armature inductance, which is preferable. The inner-magnet type can have more magnetic flux, and decreases demagnetization effects due to armature reaction. Recently, qualities of permanent magnets have been very much improved. Ferrite magnets and samarium–cobalt magnets are now in use.

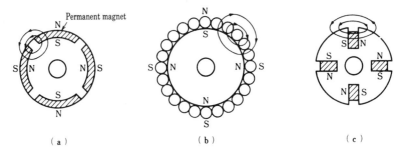

Fig. 7.1 Rotors of the permanent-magnet-excited synchronous motor

The permanent-magnet-excited synchronous motor does not usually have a damper winding. Starting is performed under a state of synchronism, by changing the power supply frequency. Thus a damper winding, as a starting winding, is not necessary. Armature currents are fed and commutated by an inverter, detecting rotor position to keep synchronism. There is no possibility of stepping out and thus there is no need for a damper winding to prevent this.

In the following analysis, a permanent-magnet-excited synchronous motor is treated. However, the following analysis is also valid for a motor with a field winding, when it is is excited with constant DC current.

7.2 Circuit equation and equivalent circuit of the synchronous motor

In Fig. 7.2 is shown a model for analysing the synchronous motor. There are three-phase windings on the stator and a two-pole permanent magnet on the rotor. The rotor has no damper winding and its iron core is also laminated. Thus there is no current flow in the rotor.

The spiral vector method will be used in the synchronous motor analysis. It will lead to phase segregation, which simplifies the analysis, as it did for the induction motor. The starting equation for the motor model in Fig. 7.2 is

$$v_a = R_1 i_a + l_1 p i_a + p \lambda_{ga}. \tag{7.1}$$

This is the same fundamental equation as equation (3.6) for the induction motor. The flux linkage λ_{ga} of phase a of the armature winding is given by

$$\lambda_{ga} = M i_a + M i_b \cos \tfrac{2}{3}\pi + M i_c \cos(-\tfrac{2}{3}\pi) - \lambda\, e^{j\theta}, \tag{7.2}$$

where the last term $\lambda\, e^{j\theta}$ is the magnetic flux linkage coming from the permanent magnet expressed as a spiral vector, and θ is the spatial phase angle between the magnetic pole and the phase a winding, as shown in

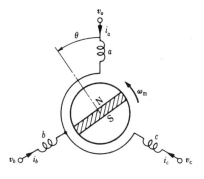

Fig. 7.2   Model of the permanent-magnet-excited synchronous motor

Fig. 7.2, and is given by

$$\theta = \omega_m t + \phi_\lambda = \omega t + \phi_\lambda. \tag{7.3}$$

Here, $\omega_m = \omega$ is in electrical radians/second.

When the three-phase armature currents have no zero-sequence component, they satisfy

$$i_a + i_b + i_c = 0, \tag{7.4}$$

and equation (7.2) becomes

$$\lambda_{ga} = \tfrac{3}{2}M i_a - \lambda \, e^{j\theta}. \tag{7.5}$$

Inserting this equation into equation (7.1), we have

$$v_a = R_1 i_a + (l_1 + \tfrac{3}{2}M)p i_a - j\omega\lambda \, e^{j(\omega t + \phi_\lambda)}. \tag{7.6}$$

This equation contains phase a only, phases b and c being left out. Similar equations are obtained for phases b and c. However, under symmetrical operation, the state variables of the three phases are symmetrical, differing only in phase angle by $\pm\tfrac{2}{3}\pi$. Thus equation (7.6) can be representative of the three phases. This is the principle of phase segregation, which was used extensively in the induction motor analysis in the preceding chapters. The necessary condition for phase segregation in this case is equation (7.4). To make equation (7.6) represent the primary three phases, subscript a is changed to 1, and then equation (7.6) becomes

$$v_1 = R_1 i_1 + L_s p i_1 + e_1. \tag{7.7}$$

Here,

$$L_s = l_1 + \tfrac{3}{2}M \tag{7.8}$$

is called the synchronous inductance, and the speed voltage or internal induced voltage e_1 is given by

$$e_1 = -j\omega\lambda \, e^{j(\omega t + \phi_\lambda)} = \omega\lambda \, e^{j(\omega t + \phi_\lambda - \frac{1}{2}\pi)} = \sqrt{2}|\dot{E}_1| \, e^{j(\omega t + \phi_i)}. \tag{7.9}$$

Here, $|\dot{E}_1| = \omega\lambda/\sqrt{2}$ and $\phi_i = \phi_\lambda - \pi/2$.

The corresponding equivalent circuit is shown in Fig. 7.3. The output p_{10} and torque t_1 per phase are

$$p_{10} = \mathrm{Re}[e_1]\,\mathrm{Re}[i_1], \tag{7.10}$$

$$t_1 = \frac{P}{2\omega}p_{10} \quad [\text{N m}] \tag{7.11}$$

The three-phase output and torque are not three times those of equations

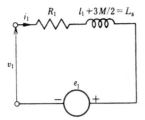

Fig. 7.3 Transient-state equivalent circuit of the permanent-magnet-excited synchronous motor (without damper)

(7.10) and (7.11) but are respectively given by

$$p_0 = 3 \, \mathrm{Re}[e_1 i_1^*], \tag{7.12}$$

$$t_3 = \frac{P}{2\omega} p_0 = \tfrac{3}{2} P \, \mathrm{Im}[\lambda \, e^{j\theta} i_1^*] \qquad [\mathrm{N \, m}], \tag{7.13}$$

where * indicates the complex conjugate.

7.3 Steady-state characteristics of the synchronous motor

Under steady-state operation, state variables expressed as spiral vectors become circular vectors and equation (7.7) becomes

$$\sqrt{2}\dot{V}_1 = R_1\sqrt{2}\dot{I}_1 + L_s p(\sqrt{2}\dot{I}_1) + \sqrt{2}\dot{E}_1,$$

whence we get the following circular vector equation:

$$\dot{V}_1 = R_1\dot{I}_1 + j\omega L_s\dot{I}_1 + \dot{E}_1. \tag{7.14}$$

Figure 7.4 shows the corresponding vector diagram for $t = 0$, which gives

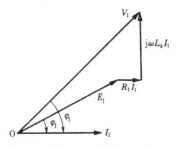

Fig. 7.4 Vector diagram of the synchronous motor

the steady-state output and torque as

$$p_0 = 3|\dot{E}_1|\,|\dot{I}_1|\cos\phi_1 \tag{7.15}$$

$$T = p_0 P/2\omega \quad (P:\text{number of poles}) \quad [\text{N m}] \tag{7.16}$$

The speed voltage or internal induced voltage $|\dot{E}_1|$ is given by equation (7.9) as

$$|\dot{E}_1| = \omega\lambda/\sqrt{2}. \tag{7.17}$$

From equations (7.15)–(7.17), we can write the torque as

$$T = \left(\frac{3P}{2\sqrt{2}}\right)\lambda|\dot{I}_1|\cos\phi_1. \tag{7.18}$$

If the power factor $\cos\phi_1$ is kept constant, then T is proportional to $|\dot{I}_1|$, as shown in Fig. 7.5. This proportional relation between T and $|\dot{I}_1|$ is a desirable relation in a control motor. When I_1 and E_1 are in phase, as shown in Fig. 7.6, $\cos\phi_1 = 1$ and the armature current $|\dot{I}_1|$ is minimized for a given torque value. In order to perform current-minimum control, the phase angle $\angle E_1$ must be sensed. This can be done by sensing the rotor position, since magnetic poles are fixed on the rotor, as shown in Fig. 7.7. When the centre line of the north pole coincides with the centre line of the phase a winding, the instantaneous value of the speed voltage E_1 becomes zero, and at this moment the instantaneous value of the phase a current I_a becomes zero and then is switched to the opposite sign. Then $\cos\phi_1$ becomes 1. A synchronous

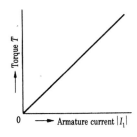

Fig. 7.5 Torque versus armature current curve of the synchronous motor

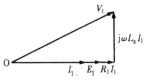

Fig. 7.6 Vector diagram of the synchronous motor (I_1 and E_1 are in phase)

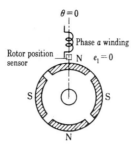

Fig. 7.7   Phase relation between rotor position and speed voltage E_1

motor with this type of armature current commutation is sometimes called
a brushless DC motor. Under this current commutation of armature
currents, the power factor is kept at 1 and there is no possibility of stepping
out.

Torque can also be controlled by terminal voltage. First, the armature
current $|\dot{I}_1|$ is determined as proportional to torque. Then, the terminal
voltage $\dot{V}_1$ is given by

$$\dot{V}_1 = \dot{E}_1 + (R_1 + j\omega L_s)\dot{I}_1. \tag{7.19}$$

It is desirable for E_1 and I_1 to be in phase, as shown in Fig. 7.6. With the
terminal voltage thus determined, the desired torque is generated.

When the power supply frequency ω is not small and R_1 is negligible
compared with ωL_s, equation (7.19) becomes

$$\dot{V}_1 = \dot{E}_1 + j\omega L_s \dot{I}_1, \tag{7.20}$$

and the following well-known relation holds:

$$T = \frac{3P|V_1|\,|E_1|}{2\omega^2 L_s} \sin(\phi_t - \phi_i) = \frac{3P|V_1|\lambda}{2\sqrt{2}\omega L_s} \sin(\phi_t - \phi_i) \quad [\text{N m}], \tag{7.21}$$

where $\phi_t - \phi_i$ is the phase difference between V_1 and E_s and is called the
power angle or internal phase angle. If this power angle is kept constant,
the torque is proportional to $|V_1|/\omega$.

7.4   Transient analysis of the synchronous motor

7.4.1   *Transient analysis of the synchronous motor without damper winding*

As explained in Section 7.2, a control synchronous motor of the permanent-
magnet type usually has no damper winding, and the rotor iron core is

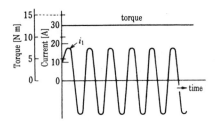

Fig. 7.8   Transient response of current-input control of the synchronous motor

usually laminated to eliminate eddy current loss. Consequently, there is no current flowing in the rotor.

Equation (7.7) and the equivalent circuit in Fig. 7.3 are also valid for transient-state analysis. This is a one-mesh circuit, and so, if the armature current is given as the control input, there cannot occur any electromagnetic transient phenomena. Therefore, current and torque are instantaneous in response to the current-input control, as shown in Fig. 7.8.

There is one problem with the sudden rise of armature current, as shown in Fig. 7.8. At $t = 0$, the armature current must rise vertically by 12 A. It is theoretically impossible for the armature current to rise discontinuously through a large synchronous inductance L_s. Actually, though, the current rise di_1/dt is kept within a certain limit to reduce the impulsive voltage $L_s \, di_1/dt$. So L_s must be reduced to reduce the impulsive voltage. The surface-permanent-magnet-type motors shown in Fig. 7.1(a, b) are effective in reducing L_s.

Next, transient phenomena in voltage-input control of synchronous motor torque will be analysed. Equation (7.7) is valid for this case also. It is assumed that the motor is kept in synchronism, and hence V_1 and e_1 have the same frequency. Thus the homogeneous equation of equation (7.7) is

$$R_1 i_1 + L_s p i_1 = 0. \tag{7.22}$$

The general transient solution is

$$i_{1t} = A \, e^{-\lambda t}, \qquad \lambda = R_1/L_s = R_1/(l_1 + \tfrac{3}{2}M). \tag{7.23}$$

The steady-state armature current is expressed by the following circular vector:

$$i_{1s} = \sqrt{2}|\dot{I}_1| \, e^{j(\omega t + \phi_1)} = 2\sqrt{\dot{I}_1}. \tag{7.24}$$

Here, the circular vector $\dot{I}_1$ is given by

$$\dot{I}_1 = \frac{\dot{V} - \dot{E}_1}{R_1 + j\omega L_s}. \tag{7.25}$$

The general solution is

$$i_1 = i_{1t} + i_{1s} = A\,e^{-\lambda t} + \sqrt{2}|\dot{I}_1|\,e^{j(\omega t + \phi_1)}. \tag{7.26}$$

When the motor is running under no load, and a control voltage is suddenly applied, the initial condition is $i_1 = 0$ at $t = 0$ in equation (7.26). Then we get

$$A = -\sqrt{2}|\dot{I}_1|\,e^{j\phi_1}. \tag{7.27}$$

Inserting this in equation (7.26), we get

$$i_1 = \sqrt{2}|\dot{I}_1|\,e^{j(\omega t + \phi_1)} - \sqrt{2}|\dot{I}_1|\,e^{j\phi_1}\,e^{-\lambda t}. \tag{7.28}$$

The torque is given by (see equations (7.12) and (7.13))

$$t_3 = \frac{3P}{4\omega}\mathrm{Re}[ei_1^*] = \frac{3P}{2\sqrt{2}}\lambda|\dot{I}_1|[\cos(\phi_i - \phi_1) - e^{-\lambda t}\cos(\omega t + \phi_i - \phi_1)]. \tag{7.29}$$

Table 7.1 gives typical rated values and constants for a synchronous motor. While it is running under no load voltage, control of rated voltage is suddenly applied. Transient current and torque were calculated using equations (7.28) and (7.29) and the results are shown in Figs 7.9 and 7.10. The decay time constant $1/\lambda$ is 0.0573 s, which is rather large. Although the transient of each phase is dependent on the switching phase, the three-phase torque is not. Voltage-input control of synchronous motor torque is rather slow in response and therefore not practical.

7.4.2 *Transient analysis of the synchronous motor with damper winding*

As explained in the preceding section, permanent-magnet-excited synchronous motors do not usually have a damper winding. In contrast, most large synchronous motors have both a damper winding and a field winding. When there are windings on the rotor, analysis of the motor becomes a little more complicated. The field winding is usually a single-phase winding, which makes motor analysis difficult. If the field current is a constant DC current, the flux linkage produced by it is also constant and its analysis becomes the same as for the permanent-magnet-excited motor. This is true, irrespective of whether it has a damper winding or not.

Table 7.1 Rating and circuit constants of a synchronous motor

Rating	3.7 kW, 200 V, 50 Hz, 4 poles
Circuit constants	$R_1 = 0.283\ \Omega$, $R_2 = 0.367\ \Omega$, $x_1 = 0.153\ \Omega$, $x_2 = 0.113\ \Omega$, $x_m = \frac{3}{2}M\omega = 4.943\ \Omega$

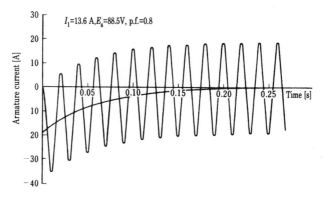

Fig. 7.9 Transient current for voltage-input control of the synchronous motor without damper.

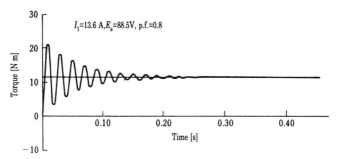

Fig. 7.10   Torque transient of the synchronous motor without damper

Figure 7.11 shows the synchronous motor model for the analysis which follows. It has a permanent magnet for excitation and a damper winding. The damper winding is usually a cage winding, but it is represented by the equivalent three-phase winding, as was done for the analysis of the induction motor in this book.

The spiral vector representation will be used and it will lead to the phase segregation analysis, as was the case for the induction motor. The starting circuit equation for phase a is

$$v_a = R_1 i_a + l_1 p i_a + p \lambda_{ga}, \tag{7.30}$$

which is the same as equation (7.1). However, the flux linkage λ_{ga} is different from that in equation (7.2) and is given by

$$\lambda_{ga} = Mi_a + Mi_b \cos \tfrac{2}{3}\pi + Mi_c \cos(-\tfrac{2}{3}\pi)$$
$$+ Mi_r \cos\theta + Mi_s \cos(\theta + \tfrac{2}{3}\pi) + Mi_t \cos(\theta - \tfrac{2}{3}\pi) - \lambda\, e^{j\theta}. \tag{7.31}$$

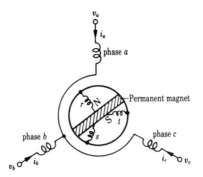

Fig. 7.11 Model for analysis of the synchronous motor with damper

The last term is the flux linkage due to the permanent magnet. When motor operation is symmetrical, the following equations holds:

$$i_a + i_b + i_c = 0, \qquad i_r + i_s + i_t = 0, \tag{7.32}$$

irrespective of whether the motor is in the steady state or transient state. And when state variables are expressed as spiral vectors, the following equations hold:

$$i_b - i_c = -j\sqrt{3}i_a, \qquad i_s - i_t = -j\sqrt{3}i_r, \tag{7.33}$$

as shown in equations (3.13) and (3.14) for the induction motor. Then flux linkage of equation (7.31) is modified as follows:

$$\begin{aligned}
\lambda_{ga} &= M[i_a - \tfrac{1}{2}(i_b + i_c)] + M[i_r - \tfrac{1}{2}(i_s + i_t)]\cos\theta + \tfrac{1}{2}\sqrt{3}(-i_s + i_t)\sin\theta - \lambda\, e^{j\omega_m t} \\
&= \tfrac{3}{2}Mi_a + \tfrac{3}{2}Mi_r(\cos\theta + j\sin\theta) - \lambda\, e^{j\omega_m t} \\
&= \tfrac{3}{2}M(i_a + i_r e^{j\omega_m t}) - \lambda\, e^{j\omega_m t}. \tag{7.34}
\end{aligned}$$

Inserting this in equation (7.30), we get

$$\begin{aligned}
v_a &= R_1 i_a + (l_1 + \tfrac{3}{2}M)pi_a + \tfrac{3}{2}Mp(i_r e^{j\omega_m t}) - p\lambda\, e^{j\omega_m t} \\
&= R_1 i_a + (l_1 + \tfrac{3}{2}M)pi_a + \tfrac{3}{2}Mpi_r' - j\omega_m \lambda\, e^{j\omega_m t}, \tag{7.35}
\end{aligned}$$

where

$$i_r' = i_r\, e^{j\omega_m t}. \tag{7.36}$$

Equation (7.35) contains phases a and r only, these being segregated from the other phases. Since phases a and r are representative of the primary and secondary three phases respectively, subscripts a and r are changed to 1 and 2, and equation (7.35) now becomes

$$v_1 = R_1 i_1 + (l_1 + \tfrac{3}{2}M)pi_1 + \tfrac{3}{2}Mpi_2 + e_1, \tag{7.37}$$

where i_2 is

$$i_2 = i'_r = i_r e^{j\omega_m t} \tag{7.38}$$

and the speed voltage e_1 is

$$e_1 = -j\omega_m \lambda\, e^{j\omega_m t}. \tag{7.39}$$

The circuit equation of phase r of the damper winding is given by

$$0 = R_2 i_r + l_2 p i_r + p\lambda_{gr}. \tag{7.40}$$

This is the same as equation (3.7) for the induction motor. The flux linkage λ_{gr} is given by

$$\lambda_{gr} = M i_r + M i_s \cos \tfrac{2}{3}\pi + M i_i \cos(-\tfrac{2}{3}\pi)$$
$$+ M i_a \cos(-\theta) + M i_b \cos(-\theta + \tfrac{2}{3}\pi) + M i_c \cos(-\theta - \tfrac{2}{3}\pi) - \lambda. \tag{7.41}$$

This can be modified as

$$\lambda_{gr} = \tfrac{3}{2} M (i_r + i_a e^{-j\omega_m t}) - \lambda. \tag{7.42}$$

The modification process is the same as for equation (7.34). Inserting this result in equation (7.40), we have

$$0 = R_2 i_r + (l_2 + \tfrac{3}{2}M) p i_r + \tfrac{3}{2} M (e^{-j\omega_m t} p i_a - j\omega_m e^{-j\omega_m t} i_a). \tag{7.43}$$

Multiplying each term by $e^{j\omega_m t}$, we get

$$0 = R_2 i_r e^{j\omega_m t} + (l_2 + \tfrac{3}{2}M)\, e^{j\omega_m t} p i_r + \tfrac{3}{2} M (p - j\omega_m) i_a. \tag{7.44}$$

From equation (7.36), we get

$$p i'_r = j\omega_m i'_r + e^{j\omega_m t} p i_r. \tag{7.45}$$

When this is inserted into equation (7.44), we have

$$0 = R_2 i'_r + (l_2 + \tfrac{3}{2}M)(p - j\omega_m) i'_r + \tfrac{3}{2} M (p - j\omega_m) i_a. \tag{7.46}$$

If we change subscript a to 1 and i'_r to i_2, equation (7.46) becomes

$$0 = R_2 i_2 + (l_2 + \tfrac{3}{2}M)(p - j\omega_m) i_2 + \tfrac{3}{2} M (p - j\omega_m) i_1. \tag{7.47}$$

This equation and equation (7.37) are the circuit equations for the armature winding and damper winding, respectively. They are combined into the following matrix equation, where $\omega_m = \omega$ because the motor is synchronous.

$$\begin{bmatrix} v_1 - e_1 \\ 0 \end{bmatrix} = \begin{bmatrix} R_1 + (l_1 + \tfrac{3}{2}M)p & \tfrac{3}{2}Mp \\ \tfrac{3}{2}M(p - j\omega) & R_2 + (l_2 + \tfrac{3}{2}M)(p - j\omega) \end{bmatrix} \begin{bmatrix} i_1 \\ i_2 \end{bmatrix}. \tag{7.48}$$

Figure 7.12 shows the corresponding equivalent circuit, in which e_1 is the speed voltage for the synchronous motor action and e_i is the speed voltage

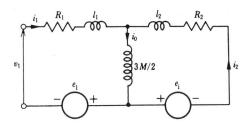

Fig. 7.12 T type transient-state equivalent circuit of the synchronous motor with damper

for the induction motor action, with e_1 given by equation (7.39) and e_i by

$$e_i = -\tfrac{3}{2}j\omega M i_1 - j\omega(l_2 + \tfrac{3}{2}M)i_2. \tag{7.49}$$

The output and torque *per phase* are as follows.
   For synchronous motor action:

$$\text{output} \quad p_{s1} = \text{Re}[e_1]\,\text{Re}[i_1], \tag{7.50}$$

$$\text{torque} \quad t_{s1} = \frac{P}{2\omega}\,p_{s1} \quad [\text{N m}]. \tag{7.51}$$

For induction motor action:

$$\text{output} \quad p_{i1} = \text{Re}[e_i]\,\text{Re}[i_2], \tag{7.52}$$

$$\text{torque} \quad t_{i1} = \frac{P}{2\omega}\,p_{i1} \quad [\text{N m}]. \tag{7.53}$$

The *total* output and torque are not $3p_{s1} + 3p_{i1}$ and $3t_{s1} + 3t_{i1}$, but are as follows:

$$\text{output} \quad p_3 = 3\,\text{Re}[e_1 i_1^*] + 3\,\text{Re}[e_i i_2^*], \tag{7.54}$$

$$\text{torque} \quad t_3 = \frac{3P}{2\omega}\,p_3 \quad [\text{N m}]. \tag{7.55}$$

The current-input control of a synchronous motor with a damper winding will be now analysed. In this case, the armature current i_1 is the control input and is given by

$$i_1 = \sqrt{2}|\dot{I}_1|e^{j(\omega t + \phi_1)}, \tag{7.56}$$

which is a circular vector. The circuit equation for this case is the second equation of the matrix equation (7.48), which is rewritten as

$$0 = \tfrac{3}{2}M(p - j\omega)i_1 + R_2 i_2 + (l_2 + \tfrac{3}{2}M)(p - j\omega)i_2. \tag{7.57}$$

Its characteristic equation is

$$R_2 + (l_2 + \tfrac{3}{2}M)(p - j\omega) = 0, \tag{7.58}$$

which gives the following characteristic root:

$$p = -\frac{R_2}{l_2 + \tfrac{3}{2}M} + j\omega. \tag{7.59}$$

The general solution is

$$i_2 = A \exp\left[\left(-\frac{R_2}{l_2 + \tfrac{3}{2}M} + j\omega\right)t\right]. \tag{7.60}$$

Here, A is an arbitrary constant, which is to be determined by an initial condition. The steady-state term is not present here, because the steady-state secondary current becomes zero by inserting equation (7.56) into equation (7.57).

To the synchronous motor running under no load, the armature current of equation (7.56) is suddenly impressed as the control input. The torque command determines $|\dot{I}_1|$ according to equation (7.18) and the rotor position sensor shown in Fig. 7.7 reads ϕ_1. Thus $|\dot{I}_1| \, \mathrm{e}^{j\phi_1}$ is determined. Before $t = 0$, the damper current $i_2 = 0$. At $t = 0$, the armature current i_1 of equation (7.56) rises suddenly to $\sqrt{2}|\dot{I}_1| \, \mathrm{e}^{j\phi_1}$. Subsequently, the damper current i_2 rises suddenly to

$$i_2(0+) = -\frac{\tfrac{3}{2}M}{l_2 + \tfrac{3}{2}M} \sqrt{2}|\dot{I}_1| \, \mathrm{e}^{j\phi_1}. \tag{7.61}$$

Inserting this initial condition in equation (7.60), we get

$$A = -\frac{\tfrac{3}{2}M}{l_2 + \tfrac{3}{2}M} \sqrt{2}|\dot{I}_1| \, \mathrm{e}^{j\phi_1}, \tag{7.62}$$

and inserting this in equation (7.60), we get

$$i_2 = -\frac{\tfrac{3}{2}M}{l_2 + \tfrac{3}{2}M} \sqrt{2}|\dot{I}_1| \exp\left(-\frac{R_2}{l_2 + \tfrac{3}{2}M} t\right) \mathrm{e}^{j(\omega t + \phi_1)}. \tag{7.63}$$

We shall now obtain the torque solution. For synchronous motor action, the torque per phase is

$$\begin{aligned}
t_{s1} &= \frac{P}{2\omega} \operatorname{Re}[e_1] \operatorname{Re}[i_1] \\
&= \frac{P}{2\omega} \{\omega\lambda \sin(\omega t + \phi_i)[\sqrt{2}|\dot{I}_1| \cos(\omega t + \phi_1)]\} \\
&= \frac{P\lambda|\dot{I}_1|}{2\sqrt{2}} [\sin(\phi_i - \phi_1) + \sin(2\omega t + \phi_i + \phi_1)]. \tag{7.64}
\end{aligned}$$

Here, e_1 of equation (7.39) is rewritten as

$$e_1 = -j\omega\lambda \, e^{j(\omega t + \phi_i)}. \tag{7.65}$$

Phase angles ϕ_i and ϕ_1 have a phase difference of $\frac{2}{3}\pi$ among three phases, and so the second term of equation (7.64) cancels out when the three phases are added together. Thus the three-phase torque is

$$t_{s3} = \frac{3P\lambda|\dot{I}_1|}{2\sqrt{2}} \sin(\phi_i - \phi_1), \tag{7.66}$$

which has no transient term.

The torque per phase for induction motor action is given by equation (7.53) as

$$t_{i1} = \frac{P}{2\omega} \operatorname{Re}[e_i] \operatorname{Re}[i_2]$$

$$= \frac{P}{2\omega} \operatorname{Re}\left[-\tfrac{3}{2}j\omega M i_1 - j\omega(l_2 + \tfrac{3}{2}M)i_2\right]$$

$$\times \operatorname{Re}\left[-\frac{\tfrac{3}{2}M}{l_2 + \tfrac{3}{2}M}\sqrt{2}|\dot{I}_1|\exp\left(-\frac{R_2}{l_2 + \tfrac{3}{2}M}t\right)e^{j(\omega t + \phi_1)}\right]$$

$$= \tfrac{1}{2}P\left[\tfrac{3}{2}M\sqrt{2}|\dot{I}_1|\sin(\omega t + \phi_1)\right]\left[1 - \exp\left(-\frac{R_2}{l_2 + \tfrac{3}{2}M}t\right)\right]$$

$$\times \left[-\frac{\tfrac{3}{2}M}{l_2 + \tfrac{3}{2}M}\sqrt{2}|\dot{I}_1|\exp\left(-\frac{R_2}{l_2 + \tfrac{3}{2}M}t\right)\cos(\omega t + \phi_1)\right]$$

$$= \tfrac{3}{4}PM\frac{\tfrac{3}{2}M}{l_2 + \tfrac{3}{2}M}|\dot{I}_1|^2\left[1 - \exp\left(-\frac{R_2}{l_2 + \tfrac{3}{2}M}t\right)\right]\sin(2\omega t + 2\phi_1). \tag{7.67}$$

In this equation, ϕ_1 has a phase difference of $\frac{2}{3}\pi$ among the three phases, and thus when the three phases are added together, the three-phase torque becomes

$$t_{i3} = 0. \tag{7.68}$$

Transient current flows in the damper winding, which does not produce any torque if the motor is kept synchronous. The total torque thus becomes

$$t_3 = t_{s3} + t_{i3} = t_{s3} = \frac{3P\lambda|I_1|}{2\sqrt{2}} \sin(\phi_i - \phi_1), \tag{7.69}$$

and the torque response of current-input control has no transient and is instantaneous.

As explained above, the damper winding carries transient current, but produces no torque. Thus it is thought that the damper is not useful, and is not usually provided on the control synchronous motor. However, it is very useful in reducing the effective armature inductance, which is very large and a nuisance in fast torque control.

Now the voltage-input control of a synchronous motor with a damper winding will be analysed. The characteristic equation of equation (7.48) becomes

$$
\begin{vmatrix}
R_1 + (l_2 + \tfrac{3}{2}M)p & \tfrac{3}{2}Mp \\
\tfrac{3}{2}M(p - j\omega) & R_2 + (l_2 + \tfrac{3}{2}M)(p - j\omega)
\end{vmatrix} = 0. \qquad (7.70)
$$

This is of second order in p and has two roots. Let these be denoted by

$$
\delta_1 = -\frac{1}{T_1} + j\omega_1, \qquad \delta_2 = -\frac{1}{T_2} + j\omega_2. \qquad (7.71)
$$

As a result, the general solution of equation (7.48) is given by

$$
i_1 = A_1\, e^{-t/T_1}\, e^{j\omega_1 t} + A_2\, e^{-t/T_2}\, e^{j\omega_2 t} + \sqrt{2}|\dot{I}_1|\, e^{j(\omega t + \phi_1)}. \qquad (7.72)
$$

The third term is the steady-state armature current. The corresponding damper current is given by the second equation of matrix equation (7.48), which, inserting i_1 from equation (7.72), gives

$$
i_2 = -\frac{\tfrac{3}{2}M(\delta_1 - j\omega)}{R_2 + (l_2 + \tfrac{3}{2}M)(\delta_1 - j\omega)}\, A_1 e^{-t/T_1}\, e^{j\omega_1 t}
$$
$$
-\frac{\tfrac{3}{2}M(\delta_2 - j\omega)}{R_2 + (l_2 + \tfrac{3}{2}M)(\delta_2 - j\omega)}\, A_2\, e^{-t/T_2}\, e^{j\omega_2 t} \qquad (7.73)
$$

No steady-state current flows in the damper winding.

When the motor is running under no voltage, the initial conditions are $i_1 = 0$ and $i_2 = 0$ at $t = 0$. Inserting these, we obtain the arbitrary constants A_1 and A_2 of equations (7.72) and (7.73) as follows:

$$
A_1 = \frac{(\delta_2 - j\omega)[R_2 + L_2(\delta_1 - j\omega)]}{R_2(\delta_1 - \delta_2)}\sqrt{2}|\dot{I}_1|\, e^{j\phi_1}, \qquad (7.74)
$$

$$
A_2 = \frac{(\delta_1 - j\omega)[R_2 + L_2(\delta_2 - j\omega)]}{R_2(\delta_1 - \delta_2)}\sqrt{2}|\dot{I}_1|\, e^{j\phi_1}. \qquad (7.75)
$$

We will now obtain the three-phase torque. For synchronous motor action (see Appendix VI) the torque is

$$t_{s3} = \tfrac{3}{2}P\sqrt{2}|\dot{I}_1|\lambda\left(\frac{|\delta_2 - j\omega||R_2 + L_2(\delta_1 - j\omega)|}{2R_2|\delta_1 - \delta_2|}\, e^{-\lambda_1 t}\cos(\omega_2 t - \gamma_1 + \phi_0 - \phi_1)\right.$$

$$-\frac{|\delta_1 - j\omega||R_2 + L_2(\delta_2 - j\omega)|}{R_2|\delta_1 - \delta_2|}\, e^{-\lambda_2 t}$$

$$\left. \times\, \cos(\omega_1 t - \gamma_2 + \phi_0 - \phi_1) + \tfrac{1}{2}\cos(\phi_0 - \phi_1)\right). \qquad (7.76)$$

For induction motor action, the torque is

$$t_{i3} = \tfrac{3}{2}P\,\frac{(\tfrac{3}{2}M)^2|\delta_1 - j\omega|\,|\delta_2 - j\omega|^2}{R_2|\delta_1 - \delta_2|^2}\,|I_1|^2$$

$$\times\, \{e^{-(\lambda_1 + \lambda_2)t}\sin[(\omega_2 - \omega_1)t + \gamma_3 - \gamma_4] + e^{-2\lambda_1 t}\sin(\gamma_4 - \gamma_3)\}$$

$$+\, \tfrac{3}{2}P\,\frac{(\tfrac{3}{2}M)^2|\delta_1 - j\omega|^2|\delta_2 - j\omega|}{R_2|\delta_1 - \delta_2|^2}\,|I_1|^2$$

$$\times\, \{e^{-(\lambda_1 + \lambda_2)t}\sin[(\omega_2 - \omega_1)t + \gamma_5 - \gamma_3] + e^{-2\lambda_2 t}\sin(\gamma_5 - \gamma_3)\}$$

$$+\, \tfrac{3}{2}P\,\frac{(\tfrac{3}{2}M)^2|\delta_1 - j\omega|\,|\delta_2 - j\omega|}{R_2|\delta_1 - \delta_2|}\,|I_1|^2$$

$$\times\, [e^{-\lambda_2 t}\sin(\omega_1 t - \gamma_3) - e^{-\lambda_1 t}\sin(\omega_2 t - \gamma_3)]$$

$$\text{(see Appendix VI).} \quad (7.77)$$

Here, $\lambda_1 = 1/T_1$ and $\lambda_2 = 1/T_2$. The total three-phase torque is

$$t_3 = t_{s3} + t_{i3}. \qquad (7.78)$$

The steady-state torque exists only in t_{s3} as the last term of equation (7.76) and t_{i3} has no steady-state torque. Derivation of equations (7.76) and (7.77) are given in Appendix VI.

7.4.3   Examples of synchronous motor torque response

Having obtained the transient solutions of synchronous motor torque in the preceding sections, we now give numerical examples for the synchronous motor of Table 7.1.

As shown in Fig. 7.8, an instantaneous torque response is obtained for current-input control of the synchronous motor without a damper when the motor running under no load is suddenly impressed with a control input of armature current.

Transient solutions for the current-input control of a synchronous motor with a damper were obtained as equations (7.63) and (7.69). These equations

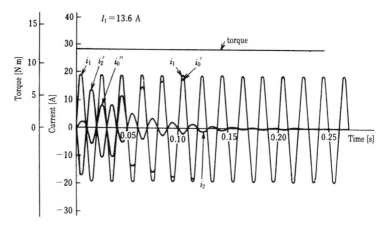

Fig. 7.13 Transient responses of currents and torque of current-input control of the synchronous motor with damper in Table 7.1

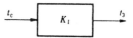

Fig. 7.14 Transfer function of current-input control of the synchronous motor with or without damper

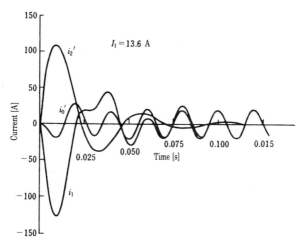

Fig. 7.15 Transient current of voltage-input control of torque for the synchronous motor with damper in Table 7.1

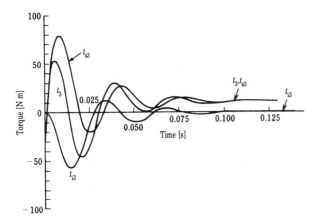

Fig. 7.16 Torque response of voltage-input control for the synchronous motor with damper in Table 7.1

were calculated for the motor in Table 7.1, and the results are shown in Fig. 7.13. Although the currents have transient, the torque does not, and its response is instantaneous.

For current-input control of synchronous motor torque, the transfer function is a real constant, as shown in Fig. 7.14, irrespectively of whether the motor has a damper or not.

Transient phenomena in a synchronous motor without a damper for voltage-input control are shown in Figs. 7.9 and 7.10. For the motor with a damper, transient solutions of voltage-input control were obtained as equations (7.72)–(7.78), which give response curves of currents and torque shown in Figs 7.15 and 7.16. Both currents and torques have severe transients and are slow in response. Thus voltage-input controls are not practical, irrespective whether the motor has a damper or not.

It should be pointed out here that the spiral vector method can be applied to analyses of the salient-pole synchronous machine, as explained in Appendix V.

8 Comparison of the induction motor and the synchronous motor as control motors

In modern motor control, control output is motor torque. By controlling the motor torque, state variables of driven equipment, such as speed, position, etc., are controlled to obtain their desired values. What is fed back in the control loop is speed, position, etc. The driving motor is located in the forward path and is under open-loop control. In the analysis of closed-loop control, which includes a driving motor in the forward path, the motor must be represented by its transfer function. Nevertheless, in many technical papers, AC motor control is treated as motor speed control. The reason for this is that electromagnetic transient phenomena have not been sufficiently analysed and it was thought that electromagnetic transients were inseparable from motor speed. In this book, it has been shown that they can be treated separately by FAM control. Thus, transfer functions have been obtained.

In terms of the analytical results explained in this book so far, the induction motor and the synchronous motor will be compared as control motors. First, what is meant by a control motor? What is a servomotor? The latter is a small motor for quick response. It is often said that the synchronous motor (the so-called brushless DC motor) is better as an AC servomotor, while the induction motor is more suitable as a large control motor. Small synchronous motors of the permanent-magnet type are replacing DC servomotors for driving machine tools, robots, etc. On the other hand, large induction motors have begun to drive high-speed elevators, railway loco-motives, and large industrial equipment, such as steel mills, etc. However, this distinction between synchronous motors as small servomotors and induction motors as large control motors seems to be transitory. It may have been thought that large control motors allow rough control, while small servomotors require precision control. This, however, is not true.

Theoretically, at least, it is not necessary to distinguish between small and large motors, both of which are called control motors in this book. The control objective of the control motor is motor torque, not motor speed. Motor speed is the controlled state variable in most classical controls.

In this book, more pages are devoted to the induction motor than to the synchronous motor. The induction motor has a secondary circuit and exciting current. Its circuit equations are more complicated, and its analysis is more difficult. It is generally believed that the existence of an exciting current degrades motor performance. Its power factor and efficiency are poorer than those of the synchronous motor. As explained in Fig. 7.6, the synchronous motor armature current $\dot{I}_1$ and speed voltage $\dot{E}_1$ are in phase. The power factor at the terminal voltage $\dot{V}_1$ is then lower than 1. To make the power factor 1, $\dot{E}_1$ must be about the same as $\dot{V}_1$ (see Fig. 7.4) and, to achieve this, the flux per pole must be increased. This calls for large permanent magnets or a large field winding, resulting in a larger synchronous motor.

The synchronous inductance L_s of a synchronous motor is generally large, and it causes difficulty in fast control of the armature current. Even in the steady state, the terminal voltage V_1 becomes very large as a result of a large voltage drop $j\omega L_s \dot{I}_1$, especially at high speed.

Thus, large synchronous inductance causes significant waste within the motor. Armature current flowing through a large inductance produces a large revolving field in the air gap, which has nothing to do with torque generation, but causes a large wasteful voltage drop.

In synchronous servomotor design, minimization of L_s is an important design objective. Reducing electric loading and increasing magnetic loading is the general guideline to be followed. But this produces a heavier motor. In the synchronous motors shown in Fig. 7.1(a, b) permanent magnets are placed on the rotor surface. This makes the effective gap length larger, resulting in smaller L_s.

The exciting inductance $\frac{3}{2}M$ of the induction motor corresponds to the armature reaction inductance $\frac{3}{2}M$ of the synchronous motor. In the induction motor, primary current flowing through the inductance $\frac{3}{2}M$ produces a large revolving magnetomotive force (mmf), but most of this is cancelled by a secondary current. The net revolving field in the air gap corresponds to an mmf of $\frac{3}{2}M(i_1 + i_2) = \frac{3}{2}Mi_0$. Thus, the large revolving field generated by the primary current induces secondary current in the rotor and produces a torque acting with the secondary current. On the other hand, in the synchronous motor, the revolving field generated by an armature current in the air gap is not utilized in torque generation and only causes a large impedance drop $j\omega L_s i_1$, and the large inductance L_s makes difficult the fast control of the armature current. Viewed this way, the synchronous motor is a wasteful motor, in which a large mmf and a large air gap field are not utilized. Overload, which may cause demagnetization of permanent magnets, is another disadvantage.

It is often said that the exciting current of the induction motor is its disadvantage. It reduces the power factor. But it circulates only between the motor and the inverter. It does not go back further along the power lines. Because of the exciting current, pre-excitation is possible, and this entirely eliminates the electromagnetic transient, resulting in instantaneous torque response.

Analytical theory of the induction motor is more complicated and has not been well established. The analysis in this book has filled the vacant area of analysis not only of the induction motor but also of the synchronous motor as control motors. In terms of the new analytical results, the relative advantages and disadvantages look somewhat different from those of the older theories. Permanent-magnet-excited synchronous motors have had a good start as AC servomotors in replacing DC servomotors. But the induction motor, which is presently used mainly as a large control motor, will be used more widely, utilizing its superior control features.

9 AC power supply for motor control

AC motor control is making rapid progress, mainly due to the advances in AC power supply. Formerly it was not easy to control AC voltage and current as vector quantities. In AC power supplies called variable-voltage constant-frequency (VVCF) or variable-voltage variable-frequency (VVVF), only one or two AC quantities are controlled. The control was a scalar control, and its application was mostly limited to AC motor speed control. Advancement of the semiconductor inverter has made it possible to control variables of the AC power supply as vector quantities. As a result, AC motor control has made remarkable progress. In the preceding chapters, the analysis and control theories of AC motors, which had been considerably lacking, have been treated and supplemented. Now AC power supply will be explained.

The recent development of the AC power supply is identical to that of the inverter technique. This is mostly due to developments of semiconductor power elements, such as power transistors and thyristors. Maximum ratings of the power transistor are now more than 1000 V and 500 A. The thyristor has reached maximum ratings of 4500 V and 3000 A. The gate-turn-off (GTO) thyristor has about the same maximum ratings as the thyristor. MOSFET power transistors and insulated gate power transistors can now handle carrier frequencies higher than 20 kHz for pulse width modulation. These advanced characteristics of semiconductor power elements have greatly improved inverter performance, with its output wave form being a good sine wave. Here, the inverter operation, in the context of AC motor control, will be briefly explained.

An example of a circuit and output voltages of the transistor inverter is shown in Fig. 9.1. Six sets of power transistors and diodes connected in parallel are connected between a DC power supply and an AC motor. When the power transistors, which act as switching elements, are turned on and off with 180° conducting periods, which are displaced by 120° in turn, the output voltages between lines are shown in Fig. 9.1. Their waveforms are of six steps, in which the fundamental components are three-phase sinusoidal waves. But the waveforms are not good sine waves and fast voltage control is not possible with them. Thus, this inverter is not apppropriate for fast control of AC motor torque.

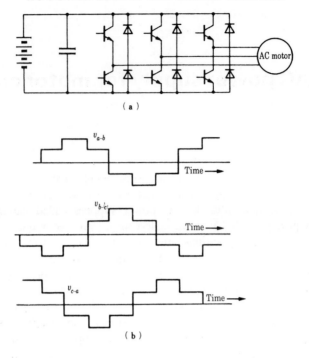

Fig. 9.1 (a) Transistor inverter circuit and (b) output voltages between lines

For a high-performance inverter, pulse modulation control must be adopted. As shown in Figs. 9.2 and 9.3, inverter output consists of a series of modulated pulses. A sine wave is derived as a fundamental component of these pulses, as shown in the figures. In Fig. 9.2, the pulse width remains unchanged, while the pulse amplitude is modulated. This is called pulse amplitude modulation (PAM). In Fig. 9.3, the pulse amplitude remains unchanged, while the pulse width is modulated in such a way that each pulse width is proportional to the instantaneous value of the desired sine wave. This is called pulse width modulation (PWM).

The fundamental structure of the PAM and PWM inverters, combined into one common circuit, is shown in Fig. 9.4, in single-phase representation for three phases. Three-phase power supply is converted to DC by a converter. Then, an inverter changes DC to three-phase AC. Since the PWM inverter is faster in control, it is more widely used than the PAM inverter. Although voltage and current always go together, one of them is chosen as a control output of the inverter. When voltage is the controlled output of the inverter, the inverter is a voltage-source inverter. When current is the controlled output, it is the current-source inverter. The current-source inverter has infinite internal impedance and the voltage-source inverter has

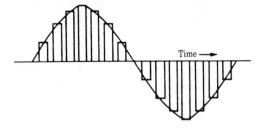

Fig. 9.2 Waveform for PAM

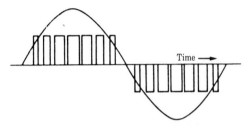

Fig. 9.3 Waveform for PWM

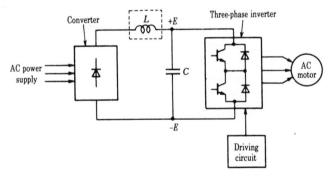

Fig. 9.4 Fundamental structure of the voltage- and current-source three-phase inverters

zero internal impedance. This theoretical difference of the internal impedances causes the different control responses in an AC motor, explained in Chapters 6 and 7.

The current-source inverter needs a large inductance L, enclosed by a broken line in Fig. 9.4. The voltage-source inverter does not need the large

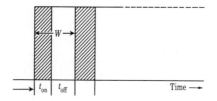

Fig. 9.5   Pulse series for the PWM inverter

inductance L but needs a large capacity C. Large C is preferable to large L, and the voltage-source inverter is preferable to the current-source inverter.

The control algorithm of the PWM three-phase inverter will now be explained. An instantaneous value of the output voltage or current of the inverter is given as a mean value of carrier frequency pulses. If the carrier frequency is sufficiently high, the waveform of the output voltage or current is a very good sine wave. The width of each pulse consists of on-time and off-time, as shown in Fig. 9.5. The instantaneous value of the output voltage or current is determined as the mean value of each pulse, which is proportional to the on-time. Thus, when the instantaneous value of the inverter output is v, the on-time is given by

$$t_{on} = Wv/E = Kv. \tag{9.1}$$

Here, W is the pulse period and $\pm E$ is the DC input voltage to the inverter, as shown in Fig. 9.4. When v is positive the output terminal of the inverter is connected to $+E$, and when negative it is connected to $-E$ through the six semiconductor elements. This is the control algorithm of the three-phase inverter. Each phase of the inverter is controlled in the same way, the only difference being the instantaneous values of the signal v, which have phase differences of $\pm\frac{2}{3}\pi$.

An example of the structure of the three-phase inverter is shown in Fig. 9.6. The unit sine wave generator generates the control signal sine wave shown at the top of Fig. 9.7. The triangle wave generator generates the triangle wave of the carrier frequency, also shown at the top of Fig. 9.7. At cross points between control sine waves and the triangle wave, the output elements are turned on or off.

Thus the on-time of equation (9.1) is obtained for each pulse, and the pulse series shown in the second, third, and fourth diagrams in Fig. 9.7 are obtained. Figure 9.8 shows an example of the output voltage and current for the PWM inverter. The AC current is a good sine wave.

For the current-source inverter, the control signal voltage v in equation (9.1) is replaced by control signal current i. DC voltage input $\pm E$ in equation (9.1) is replaced by DC current input $\pm I$. Otherwise, the control algorithm

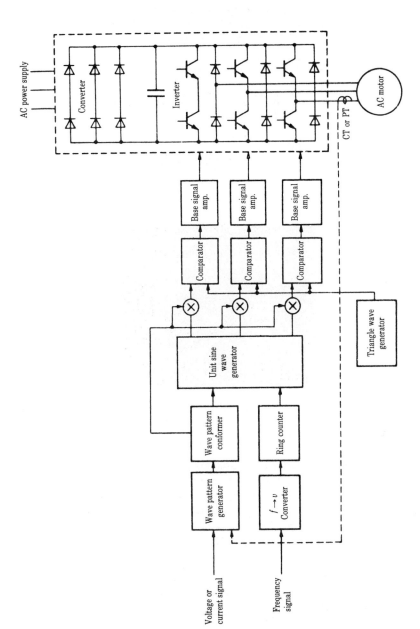

Fig. 9.6 Control circuit for the PWM inverter

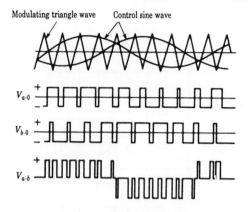

Fig. 9.7   Control signals and output voltage for the PWM inverter

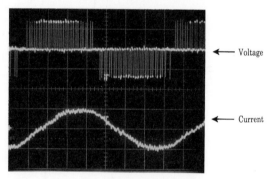

Fig. 9.8   Output voltage and current for the PWM inverter

of the three-phase current-source inverter is the same as for the voltage-source inverter.

The current-source inverter is suitable for performing current-input control of AC motors, such as T-I type FAM control of the induction motor (Fig. 3.19), while the voltage-source inverter is suitable for performing voltage-input control of AC motors, such as T-II or T type FAM control of the induction motor (Fig. 3.20). These combinations between the inverter and the AC motor are natural and they can be operated as open-loop control of the AC motor. Cross combinations between the voltage-source inverter and the current-input control of the AC motor or between the current-source inverter and the voltage-input control of the AC motor are also possible, but they require a feedback loop of the input current or input voltage of the controlled motor (see Fig. 6.5).

10 Variable transformation theories for AC motor analysis

10.1 Introduction

Sound analytical solutions of transient phenomena must precede AC motor control, otherwise an adequate control method cannot be developed. However, this area of analysis has remained vacant, and neither sound analysis nor sufficient investigation have been done, despite several existing theories of AC machines.[14] This is why most of this book is devoted to transient analysis of AC motors. The author found that conventional analytical theories were not adequate for filling the vacant area of analysis, and therefore proposed the spiral vector method. It is utilized throughout this book to analyse AC motors, and seems to have succeeded in filling the vacant area of AC motor transient analysis.

The new method is remarkable in that it uses the original variables expressed as spiral vectors, without making any variable transformations. It leads to the phase segregation method, which makes it possible to write circuit equations for both steady and transient states of three-phase machines in terms of the variables of a single phase. The general solution of the circuit equation includes both steady and transient solutions expressed as spiral vectors. The circuit equation is succinct and easy to solve. Since the variables are the original variables, the general solution can be directly connnected to the motor control computation. It has produced FAM control of induction motor torque, which has derived superior control features from the induction motor.

It seems to the author that the usefulness of the spiral-vector method is not limited to AC machine analysis, but should be extended to AC circuit analysis in general. In conventional AC circuit analysis, steady states and transient states are analysed by two separate theories, which make use of two different variable expressions, that is, phasor notation and real value expression. The spiral vector method unifies the two theories into a single theory, namely the spiral vector theory of AC circuits, where steady states and transient states are treated simultaneously, solutions for both states being included in the general solution of the circuit equation, which is expressed in terms of spiral vectors.

The spiral vector theory of AC circuits is not discussed very much in this book. However, AC motor analysis makes use of spiral vectors and there are many examples of the application of the spiral vector method throughout the book. The author hopes that the theory of stationary AC circuits will also be taught using spiral vectors in schools and universities. It does not need much modification of conventional theories written in terms of phasor notation and real value expression. Minor changes of symbols are sufficient in spite of the fact that the idea behind the changes is new.

10.2   Three-phase to two-phase transformation

The two-phase equivalent machine method has been rather popular, but it is not very useful.[14, 15] In this method, the three-phase variables of the AC machine are transformed into variables of the equivalent two-phase machine. Variables are expressed in instantaneous real values. The centre axes of the two-phase windings are spatially perpendicular to each other, and are called the d axis and q axis. This variable transformation is sometimes called d–q axis transformation. However, this name is very misleading, because the d axis and q axis are confused with the direct axis and quadrature axis of the salient poles of a synchronous machine. In synchronous machine analysis, three-phase currents are divided into a direct axis component and a quadrature axis component, without variable transformation. This is the two-reaction theory, which is different from d–q axis variable transformation.

The process of transforming three-phase variables into two-phase variables is awkward, and is not even theoretically correct without adding an additional condition. Moreover, the transformation does not have much merit in the analysis of AC machine transients. The number of phases is reduced from three to two, but the performance equation of the equivalent two-phase induction motor, for example, has a characteristic equation of the fourth degree, which is still too high and is inconvenient to use. This is one of the reasons why transient phenomena in AC machines have remained to a large extent unanalysed, and why the trans-vector or field orientation control of the induction motor has been lacking a sound theoretical basis.

It is sometimes insisted that the two-phase theory is convenient for numerical analysis of three-phase machines, but numerical solutions are not very useful in understanding transient phenomena of three-phase machines. Circuit equations written in terms of spiral vectors are also easier to solve numerically by computer (Appendix IV). For the numerical solution of AC motor performance equations, the original three-phase variables expressed in spiral vectors are also easier to use than the equivalent two-phase machine method.

10.3 Instantaneous value symmetrical component method and space vector method

The symmetrical component method was proposed by Fortescue[17] and is a powerful tool for analysing a three-phase machine under asymmetrical operation. However, the method uses the phasor representation of variables, and thus its analysis is confined to the steady state. T. Bekku[16] proposed that in the symmetrical component method, variables be expressed in instantaneous real values, in order to extend the application of the method to transient analysis of three-phase machines under asymmetrical operation. He named it the instantaneous value symmetrical component method. However, it has several problems. One is that the positive- and negative-sequence components are conjugate to each other and are not independent. Positive- and negative-sequence components contained in the original variables are not separated in the symmetrical components. Thus, this method is not compatible with Fortescue's symmetrical component method. The instantaneous value symmetrical component method did not succeed in analysing transient phenomena of AC machines and is not used in practice. These problems are clarified below.

The three-phase currents i_a, i_b, and i_c are transformed into symmetrical components i_0, i_1, and i_2 as follows:

$$\begin{bmatrix} i_0 \\ i_1 \\ i_2 \end{bmatrix} = \frac{1}{3} \begin{bmatrix} 1 & 1 & 1 \\ 1 & \alpha^2 & \alpha \\ 1 & \alpha & \alpha^2 \end{bmatrix} \begin{bmatrix} i_a \\ i_b \\ i_c \end{bmatrix} = \frac{1}{3}A \begin{bmatrix} i_a \\ i_b \\ i_c \end{bmatrix}. \tag{10.1}$$

Here, i_0 is the zero-sequence component, i_1 the positive-sequence component, and i_2 the negative-sequence component, and with respect to α the following relations hold:

$$\alpha = e^{-\frac{2}{3}j\pi}, \qquad \alpha^2 = e^{\frac{2}{3}j\pi} = \alpha^*, \qquad \alpha^3 = 1, \tag{10.2}$$

$$1 + \alpha + \alpha^2 = 0. \tag{10.3}$$

Since the three-phase currents i_a, i_b, and i_c have real values, the following relation holds in equation (10.1):

$$i_1 = i_2^*. \tag{10.4}$$

The positive- and negative-sequence components are conjugate to each other and are not independent. At this point, Bekku's symmetrical component method is different from Fortescue's, where positive- and negative-sequence components are not conjugate and are independent. Equation (10.4) makes one of them superfluous, because one is conjugate to the other.

Some examples of problems with the method are now shown. The steady-state three-phase currents contain the zero-sequence component I_0, the positive-sequence component I_1, and the negative-sequence component I_2. The instantaneous values of the three-phase currents are then given as follows:

$$\left.\begin{aligned}
i_a &= \sqrt{2}|I_0| \cos(\omega t + \phi_0) + \sqrt{2}|I_1| \cos(\omega t + \phi_1) \\
&\quad + \sqrt{2}|I_2| \cos(\omega t + \phi_2), \\
i_b &= \sqrt{2}|I_0| \cos(\omega t + \phi_0) + \sqrt{2}|I_1| \cos(\omega t + \phi_1 - \tfrac{2}{3}\pi) + \sqrt{2}|I_2| \\
&\quad \times \cos(\omega t + \phi_2 - \tfrac{2}{3}\pi), \\
i_c &= \sqrt{2}|I_0| \cos(\omega t + \phi_0) + \sqrt{2}|I_1| \cos(\omega t + \phi_1 + \tfrac{2}{3}\pi) + \sqrt{2}|I_2| \\
&\quad \times \cos(\omega t + \phi_2 - \tfrac{2}{3}\pi).
\end{aligned}\right\} \quad (10.5)$$

Transforming these currents using equation (10.1), we obtain the following symmetrical components:

$$\left.\begin{aligned}
i_0 &= \sqrt{2}|I_0| \cos(\omega t + \phi_0), \\
i_1 &= \frac{1}{\sqrt{2}}|I_1| e^{j(\omega t + \phi_1)} + \frac{1}{\sqrt{2}}|I_2| e^{-j(\omega t + \phi_2)}, \\
i_2 &= \frac{1}{\sqrt{2}}|I_1| e^{-j(\omega t + \phi_1)} + \frac{1}{\sqrt{2}} e^{j(\omega t + \phi_2)}.
\end{aligned}\right\} \quad (10.6)$$

Here, the zero-sequence component remains the same as the real value expression in equation (10.5), and the positive- and negative-sequence components become circular vectors and are conjugate to each other. Fortescue's positive- and negative-sequence components are not separated but are included together in Bekku's symmetrical components. Under the condition of the transformed variables of equation (10.6), it is difficult to proceed with an analytical treatment of AC circuits and machines. Mathematically speaking, the normalization of variables and circuit equations is not possible.

The instantaneous value symmetrical component method was proposed many years ago[16] and was introduced into the system of analytic theories.[14, 18] It did not succeed, however, and is not used in AC motor analysis at the present time of development of AC control motors.

Recently the space vector method, or space phasor method, has been proposed.[15, 19] They are the same, and start with the real value expressions for the three-sequence currents i_a, i_b, and i_c. These are transformed by the following equation:

$$i = i_a + \alpha^2 i_b + \alpha i_c. \quad (10.7)$$

This is just a positive-sequence component of the instantaneous value symmetrical component method of equation (10.1). In this method, the complex plane is superposed on a spatial plane perpendicular to the motor shaft. But this is where an explanation of the method is rather vague and mathematically incorrect. The former is an anisotropic plane, while the latter is an isotropic plane. They can be merged neither physically nor mathematically. Transformation of equation (10.7) is singular, because the number of the new variables is smaller than the number of original variables. It cannot derive the circuit equation of the AC motor mathematically.[19]

10.4 Spiral vector symmetrical component method

When the spiral vector method is applied, the asymmetrical three-phase currents of equation (10.5) are written as follows:

$$
\left.\begin{aligned}
i_a &= \sqrt{2}|\dot{I}_0|\, e^{j(\omega t + \phi_0)} + \sqrt{2}|\dot{I}_1|\, e^{j(\omega t + \phi_1)} + \sqrt{2}|\dot{I}_2|\, e^{j(\omega t + \phi_2)}, \\
i_b &= \sqrt{2}|\dot{I}_0|\, e^{j(\omega t + \phi_0)} + \sqrt{2}|\dot{I}_1|\, e^{j(\omega t + \phi_1 - \frac{2}{3}\pi)} + \sqrt{2}|\dot{I}_2|\, e^{j(\omega t + \phi_2 + \frac{2}{3}\pi)}, \\
i_c &= \sqrt{2}|\dot{I}_0|\, e^{j(\omega t + \phi_0)} + \sqrt{2}|\dot{I}_1|\, e^{j(\omega t + \phi_1 + \frac{2}{3}\pi)} + \sqrt{2}|\dot{I}_2|\, e^{j(\omega t + \phi_2 - \frac{2}{3}\pi)}.
\end{aligned}\right\} \quad (10.8)
$$

All terms are circular vectors, which represent steady states. Equation (10.1) transforms them into the following components:

$$
i_0 = \sqrt{2}|\dot{I}_0|\, e^{j(\omega t + \phi_0)}, \qquad i_1 = \sqrt{2}|\dot{I}_1|\, e^{j(\omega t + \phi_1)}, \qquad i_2 = \sqrt{2}|\dot{I}_2|\, e^{j(\omega t + \phi_2)}. \quad (10.9)
$$

These components are independent of each other, and Fortescue's symmetrical components $|\dot{I}_0|$, $|\dot{I}_1|$, and $|\dot{I}_2|$ are separated in each of the symmetrical components i_0, i_1, and i_2. Mathematically speaking, new variables are normalized, and the performance equation will also be normalized, when written in terms of them.

For the transient state, variables are expressed as spiral vectors. The circular vectors of the symmetrical components of equation (10.9) then become the following spiral vectors:

$$
i_0 = A_0\, e^{\delta_0 t}, \qquad i_1 = A_1\, e^{\delta_1 t}, \qquad i_2 = A_2\, e^{\delta_2 t}. \quad (10.10)
$$

Here, δ_1, δ_2, and δ_0 are the characteristic roots of the symmetrical components. In the spiral vector symmetrical component method, the transformation matrix is the same as Fortescue's. Thus, from the symmetrical components of equation (10.10), the three-phase currents are given as

$$
\begin{bmatrix} i_a \\ i_b \\ i_c \end{bmatrix} = 3A^{-1} \begin{bmatrix} i_0 \\ i_1 \\ i_2 \end{bmatrix} = \begin{bmatrix} 1 & 1 & 1 \\ 1 & \alpha & \alpha^2 \\ 1 & \alpha^2 & \alpha \end{bmatrix} \begin{bmatrix} i_0 \\ i_1 \\ i_2 \end{bmatrix} = A^* \begin{bmatrix} i_0 \\ i_1 \\ i_2 \end{bmatrix}. \quad (10.11)
$$

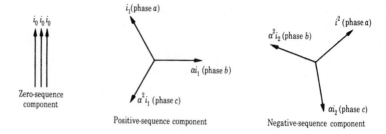

Fig. 10.1   Distribution of the symmetrical components of the spiral vectors among the three phases a, b, and c

The transformation matrix A is the same as that of equation (10.1) and is nonsingular. Thus, we have the following relation:

$$A^{-1} = \tfrac{1}{3}A^*. \tag{10.12}$$

Here, A^* is the conjugate matrix of A. Figure 10.1 shows the distribution of symmetrical components among the three phases for $t = 0$, which is the same as Fortescue's symmetrical components. But each component rotates at a different speed.

When $\delta_0 = \delta_1 = \delta_2 = j\omega$, the spiral vector symmetrical components are identical to Fortescue's symmetrical components, and the two methods are compatible.

The spiral vector symmetrical component method will now be applied to the transient analysis of the induction motor under asymmetrical operation. The starting equations for the induction motor analysis are equations (3.6) and (3.7). These equations are rewritten for three phases a, b, and c, as

$$\begin{bmatrix} v_a \\ v_b \\ v_c \end{bmatrix} = (R_1 + l_1 p) \begin{bmatrix} i_a \\ i_b \\ i_c \end{bmatrix} + p \begin{bmatrix} \lambda_{ga} \\ \lambda_{gb} \\ \lambda_{gc} \end{bmatrix}, \tag{10.13}$$

and for phases r, s, and t we have

$$\begin{bmatrix} 0 \\ 0 \\ 0 \end{bmatrix} = (R_2 + l_2 p) \begin{bmatrix} i_r \\ i_s \\ i_t \end{bmatrix} + p \begin{bmatrix} \lambda_{gr} \\ \lambda_{gs} \\ \lambda_{gt} \end{bmatrix}. \tag{10.14}$$

The magnetic flux linkages λ_{ga}, λ_{gb}, and λ_{gc} coming from the revolving gap flux are given as

$$
\begin{bmatrix} \lambda_{ga} \\ \lambda_{gb} \\ \lambda_{gc} \end{bmatrix} = M \begin{bmatrix} 1 & \cos\frac{2}{3}\pi & \cos(-\frac{2}{3}\pi) \\ \cos(-\frac{2}{3}\pi) & 1 & \cos\frac{2}{3}\pi \\ \cos\frac{2}{3}\pi & \cos(-\frac{2}{3}\pi) & 1 \end{bmatrix} \begin{bmatrix} i_a \\ i_b \\ i_c \end{bmatrix}
$$
$$
+ M \begin{bmatrix} \cos\theta & \cos(\theta+\frac{2}{3}\pi) & \cos(\theta-\frac{2}{3}\pi) \\ \cos(\theta-\frac{2}{3}\pi) & \cos\theta & \cos(\theta+\frac{2}{3}\pi) \\ \cos(\theta+\frac{2}{3}\pi) & \cos(\theta-\frac{2}{3}\pi) & \cos\theta \end{bmatrix} \begin{bmatrix} i_r \\ i_s \\ i_t \end{bmatrix}. \quad (10.15)
$$

Transforming flux linkages into symmetrical components, we have

$$
\begin{bmatrix} \lambda_{10} \\ \lambda_{11} \\ \lambda_{12} \end{bmatrix} = \tfrac{1}{3}A \begin{bmatrix} \lambda_{ga} \\ \lambda_{gb} \\ \lambda_{gc} \end{bmatrix} = \tfrac{1}{3}AM \begin{bmatrix} 1 & \cos\frac{2}{3}\pi & \cos(-\frac{2}{3}\pi) \\ \cos(-\frac{2}{3}\pi) & 1 & \cos\frac{2}{3}\pi \\ \cos\frac{2}{3}\pi & \cos(-\frac{2}{3}\pi) & 1 \end{bmatrix} A^* \begin{bmatrix} i_{10} \\ i_{11} \\ i_{12} \end{bmatrix}
$$
$$
+ \tfrac{1}{3}AM \begin{bmatrix} \cos\theta & \cos(\theta+\frac{2}{3}\pi) & \cos(\theta-\frac{2}{3}\pi) \\ \cos(\theta-\frac{2}{3}\pi) & \cos\theta & \cos(\theta+\frac{2}{3}\pi) \\ \cos(\theta+\frac{2}{3}\pi) & \cos(\theta-\frac{2}{3}\pi) & \cos\theta \end{bmatrix} A^* \begin{bmatrix} i_{20} \\ i_{21} \\ i_{22} \end{bmatrix}.
$$
$$
(10.16)
$$

Here, the symmetrical components of the primary currents are

$$
\begin{bmatrix} i_{10} \\ i_{11} \\ i_{12} \end{bmatrix} = \tfrac{1}{3}A \begin{bmatrix} i_a \\ i_b \\ i_c \end{bmatrix}, \quad (10.17)
$$

and the symmetrical components of the secondary currents are

$$
\begin{bmatrix} i_{20} \\ i_{21} \\ i_{22} \end{bmatrix} = \tfrac{1}{3}A \begin{bmatrix} i_r \\ i_s \\ i_t \end{bmatrix}. \quad (10.18)
$$

Performing the matrix multiplication, equation (10.16) becomes

$$
\begin{bmatrix} \lambda_{10} \\ \lambda_{11} \\ \lambda_{12} \end{bmatrix} = \tfrac{3}{2}M \begin{bmatrix} 0 \\ i_{11} + i_{21}\,e^{j\theta} \\ i_{12} + i_{22}\,e^{-j\theta} \end{bmatrix}. \quad (10.19)
$$

The secondary flux linkages λ_{gr}, λ_{gs}, and λ_{gt} coming from the revolving gap flux are

$$
\begin{bmatrix} \lambda_{gr} \\ \lambda_{gs} \\ \lambda_{gt} \end{bmatrix} = M \begin{bmatrix} 1 & \cos\frac{2}{3}\pi & \cos(-\frac{2}{3}\pi) \\ \cos(-\frac{2}{3}\pi) & 1 & \cos\frac{2}{3}\pi \\ \cos\frac{2}{3}\pi & \cos(-\frac{2}{3}\pi) & 1 \end{bmatrix} \begin{bmatrix} i_s \\ i_t \\ i_t \end{bmatrix}
$$

$$
+ M \begin{bmatrix} \cos(-\theta) & \cos(-\theta+\frac{2}{3}\pi) & \cos(-\theta-\frac{2}{3}\pi) \\ \cos(-\theta-\frac{2}{3}\pi) & \cos(-\theta) & \cos(-\theta+\frac{2}{3}\pi) \\ \cos(-\theta+\frac{2}{3}\pi) & \cos(-\theta-\frac{2}{3}\pi) & \cos(-\theta) \end{bmatrix} \begin{bmatrix} i_a \\ i_b \\ i_c \end{bmatrix}.
$$

$$(10.20)$$

Transforming these into symmetrical components, we get the following flux linkages:

$$
\begin{bmatrix} \lambda_{20} \\ \lambda_{21} \\ \lambda_{22} \end{bmatrix} = \tfrac{1}{3}A \begin{bmatrix} \lambda_{gr} \\ \lambda_{gr} \\ \lambda_{gt} \end{bmatrix} = \tfrac{3}{2}M \begin{bmatrix} 0 \\ i_{21} + i_{11}\, e^{-j\theta} \\ i_{22} + i_{12}\, e^{j\theta} \end{bmatrix}. \tag{10.21}
$$

Equations (10.19) and (10.21) indicate that there are no zero-sequence component flux linkages, even when there are zero-sequence component currents.

Multiplying equations (10.13) and (10.14) by $\tfrac{1}{3}A$ from the left side, we get the following two equations:

$$
\begin{bmatrix} v_{10} \\ v_{11} \\ v_{12} \end{bmatrix} = (R_1 + l_1 p) \begin{bmatrix} i_{10} \\ i_{11} \\ i_{12} \end{bmatrix} + p \begin{bmatrix} \lambda_{10} \\ \lambda_{11} \\ \lambda_{12} \end{bmatrix}, \tag{10.22}
$$

$$
\begin{bmatrix} 0 \\ 0 \\ 0 \end{bmatrix} = (R_2 + l_2 p) \begin{bmatrix} i_{20} \\ i_{21} \\ i_{22} \end{bmatrix} + p \begin{bmatrix} \lambda_{20} \\ \lambda_{21} \\ \lambda_{22} \end{bmatrix}. \tag{10.23}
$$

Inserting equations (10.19) and (10.21) into equations (10.22) and (10.23) respectively, we get the following equations.

For zero-sequence components:

$$v_{10} = (R_1 + l_1 p)i_{10}, \tag{10.24}$$

$$0 = (R_2 + l_2 p)i_{20}. \tag{10.25}$$

For positive-sequence components:

$$v_{11} = (R_1 + l_1 p)i_{11} + \tfrac{3}{2}Mp(i_{11} + e^{j\theta}i_{21}), \tag{10.26}$$

$$0 = (R_2 + l_2 p)i_{21} + \tfrac{3}{2}Mp(i_{21} + e^{-j\theta}i_{11}). \tag{10.27}$$

For negative-sequence components:

$$v_{12} = (R_1 + l_1 p)i_{12} + \tfrac{3}{2}M(i_{12} + e^{-j\theta}i_{22}), \tag{10.28}$$

$$0 = (R_2 + l_2 p)i_{22} + \tfrac{3}{2}M(i_{22} + e^{j\theta}i_{12}). \tag{10.29}$$

The following replacement is made:

$$i'_{21} = e^{j\theta}i_{21}. \tag{10.30}$$

Then equation (10.26) becomes

$$v_{11} = R_1 i_{11} + (l_1 + \tfrac{3}{2}M)p i_{11} + \tfrac{3}{2}M p i'_{21}, \tag{10.31}$$

and equation (10.27) becomes

$$0 = R_2 i_{21} + (l_2 + \tfrac{3}{2}M)p i_{21} + \tfrac{3}{2}M(e^{-j\theta}p i_{11} - j\omega_m e^{-j\theta}i_{11}). \tag{10.32}$$

If we multiply by $e^{j\theta}$, equation (10.32) becomes

$$0 = R_2 e^{j\theta}i_{21} + (l_2 + \tfrac{3}{2}M) e^{j\theta}p i_{21} + \tfrac{3}{2}M(p - j\omega_m)i_{11}. \tag{10.33}$$

From equation (10.30), we get

$$p i'_{21} = j\omega_m i'_{21} + e^{j\theta}p i_{21}. \tag{10.34}$$

Inserting equations (10.30) and (10.34) in equation (10.33), we get

$$0 = R_2 i'_{21} + (l_2 + \tfrac{3}{2}M)(p - j\omega_m)i'_{21} + \tfrac{3}{2}M(p - j\omega_m)i_{11}. \tag{10.35}$$

Equations (10.31) and (10.35) are circuit equations for the positive-sequence components and are combined in the following matrix equation:

$$\begin{bmatrix} v_{11} \\ 0 \end{bmatrix} = \begin{bmatrix} R_1 + (l_1 + \tfrac{3}{2}M)p & \tfrac{3}{2}Mp \\ \tfrac{3}{2}M(p - j\omega_m) & R_2 + (l_2 + \tfrac{3}{2}M)(p - j\omega_m) \end{bmatrix} \begin{bmatrix} i_{11} \\ i'_{21} \end{bmatrix}. \tag{10.36}$$

The corresponding equivalent circuit is shown in Fig. 10.2, in which the speed voltage e_{s1} is given by

$$e_{s1} = -j\omega_m \left[\tfrac{3}{2}M i_{11} + (l_2 + \tfrac{3}{2}M)i'_{21} \right]. \tag{10.37}$$

This circuit is called the T type transient-state equivalent circuit for positive-sequence components.

For the steady state, p becomes $j\omega$ and equation (10.36) gives the following equation:

$$\begin{bmatrix} \dot{V}_{11} \\ 0 \end{bmatrix} = \begin{bmatrix} R_1 + j(x_1 + x_m) & jx_m \\ jx_m & R_2/s + j(x_2 + x_m) \end{bmatrix} \begin{bmatrix} \dot{I}_{11} \\ \dot{I}'_{21} \end{bmatrix}. \tag{10.38}$$

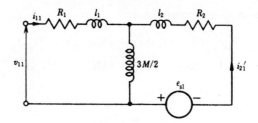

Fig. 10.2 T type transient-state equivalent circuit of the induction motor for positive-sequence components

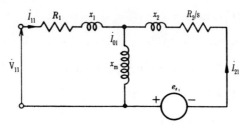

Fig. 10.3 T type steady-state equivalent circuit of the induction motor for positive-sequence components

Here, the variables are circular vectors, which are denoted by the corresponding capital letters. Figure 10.3 shows the corresponding equivalent circuit, which is the same as the T type steady-state equivalent circuit in Fig. 3.2, except that the variables are positive-sequence components expressed as circular vectors.

Negative-sequence components are treated in the same way and have the following variable replacement:

$$i'_{22} = e^{-j\theta} i_{22}. \tag{10.39}$$

Equations (10.28) and (10.29) become the following matrix equation:

$$\begin{bmatrix} v_{12} \\ 0 \end{bmatrix} = \begin{bmatrix} R_1 + (l_1 + \tfrac{3}{2}M)p & \tfrac{3}{2}Mp \\ \tfrac{3}{2}M(p + j\omega_m) & R_2 + (l_2 + \tfrac{3}{2}M)(p + j\omega_m) \end{bmatrix} \begin{bmatrix} i_{12} \\ i'_{22} \end{bmatrix}. \tag{10.40}$$

Figure 10.4 shows the corresponding equivalent circuit, in which speed voltage e_{s2} is given by

$$e_{s2} = j\omega_m [\tfrac{3}{2}M i_{12} + (l_2 + \tfrac{3}{2}M)i'_{22}]. \tag{10.41}$$

This circuit is called the T type transient-state equivalent circuit for negative-sequence components.

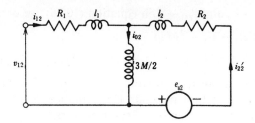

Fig. 10.4 T type transient-state equivalent circuit of the induction motor for negative-sequence components

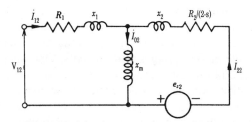

Fig. 10.5 T type steady-state equivalent circuit of the induction motor for negative-sequence components

For the steady state, the variables become circular vectors. Then p becomes $j\omega$ and equation (10.40) becomes

$$\begin{bmatrix} \dot{V}_{12} \\ 0 \end{bmatrix} = \begin{bmatrix} R_1 + j(x_1 + x_m) & jx_m \\ jx_m & R_2/s + j(x_2 + x_m) \end{bmatrix} \begin{bmatrix} \dot{I}_{12} \\ \dot{I}'_{22} \end{bmatrix}. \quad (10.42)$$

Figure 10.5 shows the corresponding equivalent circuit, which is called the T type steady-state equivalent circuit for negative-sequence components.

10.5   Spiral vector symmetrical component method and phase segregation method

When the AC motor is symmetrical in structure and its operation is also symmetrical, the zero-sequence and negative-sequence components of the voltage and current are zero. Equation (10.11) then gives

$$i_a = i_1, \quad i_b = \alpha i_1, \quad i_c = \alpha^2 i_1. \quad (10.43)$$

Here, the state variables are symmetrical spiral vectors. In the induction motor, the primary and secondary currents and voltages are identical to the corresponding positive-sequence components. Making the variable replacements

$v_{11} \to v_a$, $i_{11} \to i_a$, and $i'_{21} \to i'_r$ in equation (10.36), we obtain

$$\begin{bmatrix} v_a \\ 0 \end{bmatrix} = \begin{bmatrix} R_1 + (l_1 + \frac{3}{2}M)p & \frac{3}{2}Mp \\ \frac{3}{2}M(p - j\omega_m) & R_2 + (l_2 + \frac{3}{2}M)(p - j\omega_m) \end{bmatrix} \begin{bmatrix} i_a \\ i'_r \end{bmatrix}. \quad (10.44)$$

This equation is identical to equations (3.23) and (3.24), which were derived by the phase segregation method. This equation is valid only when the variables are expressed in terms of spiral vectors.

The AC motor transient can be analysed by the spiral vector and the phase segregation methods without making any variable transformation. This is the advantage of these methods. Using the original variables in the analysis, they are extremely convenient in the control of AC motors. These two methods provide us with FAM control of induction motor torque.

Thus, there are many examples in this book of positive-sequence components of the spiral vector symmetrical component method, which are identical to symmetrical three-phase voltages and currents. One analytical example of asymmetrical operation of the induction motor, in which there are both positive- and negative-sequence components, is now given.

Single-phase operation of a three-phase induction motor is shown in Fig. 10.6. The terminal conditions give the following relations:

$$i_a = i, \qquad i_b = -i, \qquad i_c = 0. \quad (10.45)$$

Inserting these relations in equation (10.1) gives

$$i_{10} = 0, \qquad i_{11} = \frac{1}{3}(i_a + \alpha^2 i_b) = \frac{1}{3}(1 - \alpha^2)i,$$
$$i_{12} = \frac{1}{3}(i_a + \alpha i_b) = \frac{1}{3}(1 - \alpha)i. \quad (10.46)$$

Let the terminal impedance for the positive-sequence components of the three-phase motor be denoted Z_1 and let the terminal impedance for the negative-sequence components be denoted Z_2. Then the symmetrical components of the terminal voltages are as follows:

$$v_1 = Z_1 i_{11}, \qquad v_2 = Z_2 i_{12}, \qquad v_0 = 0. \quad (10.47)$$

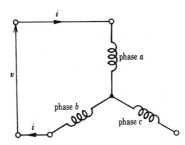

Fig. 10.6   Single-phase operation of the three-phase induction motor

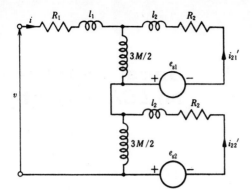

Fig. 10.7 Transient-state equivalent circuit of single-phase operation of the three-phase induction motor

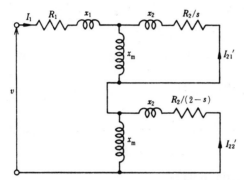

Fig. 10.8 Steady-state equivalent circuit of single-phase operation of the three-phase induction motor

The terminal voltages are

$$v_a = v_1 + v_2 = Z_1 i_{11} + Z_2 i_{12} = \tfrac{1}{3}[(1 - \alpha^2)Z_1 + (1 - \alpha)Z_2]i, \quad (10.48)$$

$$v_b = \alpha v_1 + \alpha^2 v_2 = \tfrac{1}{3}[\alpha(1 - \alpha^2)Z_1 + \alpha^2(1 - \alpha)Z_2]i. \quad (10.49)$$

The line voltage is

$$v = v_a - v_b = (Z_1 + Z_2)i. \quad (10.50)$$

The corresponding equivalent circuit for transient-state analysis is shown in Fig. 10.7. By expressing the state variables as circular vectors, for which $p = j\omega$, Fig. 10.7 becomes Fig. 10.8, which is the well-known steady-state equivalent circuit for the single-phase operation of a three-phase induction motor.

Appendix I   Spiral vector analysis of an *LCR* circuit

As an example of spiral vector analysis of the AC circuit, the *LCR* AC circuit shown in Fig. I.1 will be analysed. The circuit equation in terms of the charge q of capacitor C is given by

$$L\frac{d^2q}{dt^2} + R\frac{dq}{dt} + \frac{q}{C} = v. \tag{I.1}$$

The terminal voltage v is the AC voltage given by

$$v = \sqrt{2}|\dot{V}|\,e^{j(\omega t + \phi)} = \sqrt{2}\dot{V}, \tag{I.2}$$

which is a circular vector. Differentiating equation (I.1) with respect to time t, we obtain

$$L\frac{d^2i}{dt^2} + R\frac{di}{dt} + \frac{i}{C} = j\omega\sqrt{2}|V|\,e^{j\omega t} = j\omega\sqrt{2}\dot{V}, \tag{I.3}$$

since $dq/dt = i$. The characteristic equation for equation (I.3) is

$$Lp^2 + Rp + 1/C = 0, \tag{I.4}$$

and the characteristic roots are

$$\delta_1, \delta_2 = \frac{-R \pm (R^2 - 4L/C)^{\frac{1}{2}}}{2L}. \tag{I.5}$$

Here, the case where $R^2 - 4L/C < 0$ will be treated. The two roots are complex and conjugate, and are now denoted as

$$\delta_1, \delta_2 = -\alpha \pm j\beta, \qquad \alpha = -R/2L, \qquad \beta = (4L/C - R^2)^{\frac{1}{2}}/2L. \tag{I.6}$$

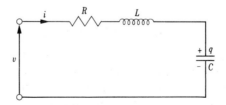

Fig. I.1   The *LCR* circuit

The general solutions of equations (I.1) and (I.3) are

$$q = A_1 \, e^{\delta_1 t} + A_2 \, e^{\delta_2 t} + \frac{\sqrt{2|\dot{I}|}}{j\omega} \, e^{j(\omega t + \phi - \theta)}, \tag{I.7}$$

$$i = \frac{dq}{dt} = A_1 \delta_1 \, e^{\delta_1 t} + A_2 \delta_2 \, e^{\delta_2 t} + \sqrt{2|\dot{I}|} \, e^{j(\omega t + \phi - \theta)}. \tag{I.8}$$

Here, $\sqrt{2|\dot{I}|} \, e^{j(\omega t + \phi - \theta)} = \sqrt{2}\dot{i}$ is the steady-state current expressed as a circular vector given by

$$i = \sqrt{2|\dot{I}|} \, e^{j(\omega t + \phi - \theta)} = \frac{j\omega\sqrt{2|\dot{V}|} \, e^{j(\omega t + \phi)}}{-\omega^2 L + j\omega R + 1/C} = \frac{j\sqrt{2|\dot{V}|} \, e^{j(\omega t + \phi + \frac{1}{2}\pi - \theta)}}{[(1/C\omega - L\omega)^2 + R^2]^{\frac{1}{2}}},$$

$$\tan \theta = \frac{R}{1/C\omega - L\omega}. \tag{I.9}$$

The arbitrary constants A_1 and A_2 in equation (I.7) are now determined by the initial conditions, which are

$$i = 0 \quad \text{and} \quad q = 0 \quad \text{at } t = 0. \tag{I.10}$$

Inserting the initial conditions in equations (I.7) and (I.8) gives

$$0 = A_1 + A_2 + \frac{\sqrt{2|\dot{I}|}}{j\omega} \, e^{j(\phi - \theta)}, \tag{I.11}$$

$$0 = (-\alpha + j\beta)A_1 + (-\alpha - j\beta)A_2 + \sqrt{2|\dot{I}|} \, e^{j(\phi - \theta)}. \tag{I.12}$$

From these equations, we get

$$A_1 = \frac{\alpha + j\beta + j\omega}{2\beta\omega} \sqrt{2|\dot{I}|} \, e^{j(\phi - \theta)}, \tag{I.13}$$

$$A_2 = \frac{\alpha - j\beta + j\omega}{2\beta\omega} \sqrt{2|\dot{I}|} \, e^{j(\phi - \theta)}. \tag{I.14}$$

Inserting these values of A_1 and A_2 in equations (I.7) and (I.8) gives

$$q = \frac{\alpha + j\beta + j\omega}{2\omega\beta} \sqrt{2|\dot{I}|} \, e^{-\alpha t} \, e^{j(\beta t + \phi - \theta)} + \frac{\alpha - j\beta - j\omega}{2\omega\beta} \sqrt{2|\dot{I}|}$$

$$\times \, e^{-\alpha t} \, e^{j(-\beta t + \phi - \theta)} + \frac{\sqrt{2|\dot{I}|}}{j\omega} \, e^{j(\omega t + \phi - \theta)}, \tag{I.15}$$

$$i = \frac{dq}{dt} = \frac{(-\alpha + j\beta + j\omega)(-\alpha + j\beta)}{2\omega\beta} \sqrt{2|\dot{I}|} \, e^{(-\alpha + j\beta)t} \, e^{j(\phi - \theta)}$$

$$+ \frac{(\alpha - j\beta - j\omega)(-\alpha - j\beta)}{2\omega\beta} \sqrt{2|\dot{I}|} \, e^{(-\alpha - j\beta)t} \, e^{j(\theta - \phi)} + \sqrt{2|\dot{I}|} \, e^{j(\omega t + \theta - \phi)}. $$

$$\tag{I.16}$$

Equations (I.15) and (I.16) give the solutions for the charge q and current i for the initial conditions $q = 0$ and $i = 0$ at $t = 0$. They are expressed as spiral vectors. The solution process is straightforward and easier than treating the solution in phasors and instantaneous real values. If instantaneous real value solutions are required, they are given by $q_{re} = \text{Re}[q]$ and $i_{re} = \text{Re}[i]$.

Appendix II   Spiral vectors and the Laplace transformation

When state variables are expressed as spiral vectors, the Laplace transformation method is easy to apply. As an example, we will solve equation (1.2) using the Laplace transformation. We write equation (1.2) as

$$a\frac{d^2 i}{dt^2} + b\frac{di}{dt} + ci = \sqrt{2}|\dot{V}|\,e^{j(\omega t + \phi)}. \tag{II.1}$$

Using the Laplace transformation, we get

$$a[s^2 I(s) - sI(0) - I'(0)] + b[sI(s) - I(0)] + cI(s) = \frac{\sqrt{2}|\dot{V}|\,e^{j\phi}}{s - j\omega}. \tag{II.2}$$

Here, $I(0)$ and $I'(0)$ are the initial conditions of i and di/dt respectively. From equation (II.2), we get

$$I(s) = \frac{\sqrt{2}|\dot{V}|\,e^{j\phi}}{(s - j\omega)(as^2 + bs + c)} + \frac{(as + b)I(0) + aI'(0)}{as^2 + bs + c}. \tag{II.3}$$

Performing the inverse Laplace transformation, we get

$$i(t) = \frac{\sqrt{2}|\dot{V}|\,e^{j\phi}}{a(j\omega)^2 + b(j\omega) + c}\,e^{j\omega t} + \frac{\sqrt{2}|\dot{V}|\,e^{j\phi}}{a(\delta_1 - j\omega)(\delta_1 - \delta_2)}\,e^{\delta_1 t}$$

$$+ \frac{\sqrt{2}|\dot{V}|\,e^{j\phi}}{a(\delta_2 - j\omega)(\delta_2 - \delta_1)}\,e^{\delta_2 t} + \frac{(a\delta_1 + b)I(0) + aI'(0)}{a(\delta_1 - \delta_2)}\,e^{\delta_1 t}$$

$$+ \frac{(a\delta_2 + b)I(0) + aI'(0)}{a(\delta_2 - \delta_1)}\,e^{\delta_2 t}. \tag{II.4}$$

Here, δ_1 and δ_2 are roots of the characteristic equation

$$as^2 + bs + c = 0. \tag{II.5}$$

When the initial values $I(0)$ and $I'(0)$ are zero, the last two terms of equation (II.4) disappear.

When spiral vectors are used, the Laplace transformation method gives a steady-state solution and a transient solution simultaneously in spiral vector form, satisfying the given initial conditions. Just one Laplace transformation formula for the exponential time function is sufficient to solve the circuit equation in most cases.

Appendix III   Transient analysis of current-input control of induction motor torque by means of the Laplace transformation

In Section 4.2.1, current-input control of induction motor torque was analysed. The circuit equation of the problem is given by equation (4.6), which is rewritten as

$$R_2 i_2 + (l_2 + \tfrac{3}{2}M)p i_2 = -\tfrac{3}{2}Mp(\sqrt{2}|\dot{I}_1| e^{j(s\omega t + \phi_1)}). \tag{III.1}$$

For the initial condition $i_2 = 0$ at $t = 0$, the Laplace transformation of equation (III.1) is

$$R_2 I_2(p) + (l_2 + \tfrac{3}{2}M)p I_2(p) = -(\tfrac{3}{2}M)\left(\sqrt{2}|\dot{I}_1| e^{j\phi_1} \frac{p}{p - js\omega}\right). \tag{III.2}$$

Here, $I_2(p)$ is the Laplace transformation of i_2, and p replaces s to avoid confusion with the slip s. Rewriting equation (III.2), we get

$$I_2(p) = -\frac{\tfrac{3}{2}M\sqrt{2}|\dot{I}_1| e^{j\phi_1}}{R_2 + (l_2 + \tfrac{3}{2}M)p} \frac{p}{p - js\omega} = -\frac{\tfrac{3}{2}M}{l_2 + \tfrac{3}{2}M} \frac{\sqrt{2}|\dot{I}_1| e^{j\phi_1}}{p + \lambda_c} \frac{p}{p - js\omega},$$

$$\lambda_c = \frac{R_2}{l_2 + \tfrac{3}{2}M}. \tag{III.3}$$

Performing the inverse Laplace transformation, we get

$$\begin{aligned}
i_2(t) &= -\frac{\tfrac{3}{2}M\sqrt{2}|\dot{I}_2| e^{j\phi_1}}{l_2 + \tfrac{3}{2}M}\left(\frac{js\omega}{js\omega + \lambda_c} e^{js\omega t} + \frac{\lambda_c}{\lambda_c + js\omega} e^{-\lambda_c t}\right) \\
&= -\frac{\tfrac{3}{2}M\sqrt{2}|\dot{I}_1| e^{j\phi_1}}{l_2 + \tfrac{3}{2}M}\left(\frac{js\omega(l_2 + \tfrac{3}{2}M)}{R_2 + js\omega(l_2 + \tfrac{3}{2}M)} e^{js\omega t} + \frac{R_2}{R_2 + js\omega(l_2 + \tfrac{3}{2}M)} e^{-\lambda_c t}\right) \\
&= -\sqrt{2}|\dot{I}_1| \frac{\tfrac{3}{2}Ms\omega}{|Z_2|} e^{j(s\omega t - \theta_2 + \phi_1 + \frac{1}{2}\pi)} - \frac{\tfrac{3}{2}M\sqrt{2}|\dot{I}_1|}{l_2 + \tfrac{3}{2}M} \frac{R_2}{|Z_2|} e^{j(\phi_1 - \theta_2)} e^{-\lambda_c t},
\end{aligned}$$

$$\tag{III.4}$$

where

$$Z_2 = R_2 + js\omega(l_2 + \tfrac{3}{2}M) = |Z_2|\, e^{j\theta_2}. \qquad \text{(III.5)}$$

The torque per phase is given by equation (4.11) as follows:

$$t_1 = \tfrac{3}{4}MP\,\mathrm{Re}[-ji_1]\,\mathrm{Re}[i_2]$$

$$= \tfrac{3}{4}MP|\dot{I}_1|^2 \left(\frac{\tfrac{3}{2}Ms\omega}{|Z_2|} [\cos\theta_2 - \cos(2s\omega t + 2\phi_1 - \theta_2)] \right.$$

$$\left. - \frac{\tfrac{3}{2}M}{l_2 + \tfrac{3}{2}M}\frac{R_2}{|Z_2|}\, e^{-\lambda_{cl}t}[\sin(s\omega t + 2\phi_1 - \theta_2) + \sin(s\omega t + \theta_2)] \right).$$

$$\text{(III.6)}$$

In this equation, ϕ_1 has a phase difference of $\pm\tfrac{2}{3}\pi$ among the three phases. The terms containing ϕ_1 in sine or cosine cancel among three phases, and the torque of three phases is

$$t_3 = \tfrac{9}{4}MP|\dot{I}_1|^2 \left(\frac{\tfrac{3}{2}Ms\omega}{|Z_2|}\cos\theta_2 - \frac{\tfrac{3}{2}M}{l_2 + \tfrac{3}{2}M}\frac{R_2}{|Z_2|}\, e^{-\lambda_{cl}t}\sin(s\omega t + \theta_2) \right). \qquad \text{(III.7)}$$

Since $\tan\theta_2 = s\omega(l_2 + \tfrac{3}{2}M)/R_2$, t_3 is zero at $t = 0$, as shown in Fig. 4.2. Torque t_3 of equation (III.7) is given by

$$t_3 = \tfrac{9}{8}MP\,\mathrm{Re}[-ji_1 i_2^*] = \tfrac{9}{8}MP\,\mathrm{Im}[i_1 i_2^*] \qquad [\mathrm{N\,m}], \qquad \text{(III.8)}$$

where $\mathrm{Im}[z]$ denotes the (real) coefficient of the imaginary part of z and * indicates the complex conjugate. Since $i_1 = i_0 - i_2$, equation (III.8) becomes

$$t_3 = \tfrac{9}{8}MP\,\mathrm{Im}[i_0 i_2^*] \qquad [\mathrm{N\,m}] \qquad \text{(III.9)}$$

The three-phase output is

$$p_{03} = \tfrac{9}{4}MP\omega_m\,\mathrm{Im}[i_1 i_2^*] = \tfrac{9}{4}MP\omega_m\,\mathrm{Im}[i_0 i_2^*], \qquad \text{(III.10)}$$

where ω_m is the speed of the motor in radians/second.

Appendix IV   Numerical solution of the performance equations for the induction motor

Derivation of the performance equation is the first step in the analysis of circuits and machines. When the analytical solution is obtained, the analysis process is almost finished, and we are ready to derive the analytical conclusions of the performance. When the performance equation cannot be solved analytically, we resort to a computer to obtain its numerical solution.

The numerical solution is less informative about the performance than the analytical solution. The numerical solution is quite often insufficient for understanding the physical nature of the phenomena concerned, as the oscillogram is not.

Therefore, all the circuit equations in this book are solved analytically and in this sense the analyses are complete and ready for practical use. By making use of the analytical results obtained, the details of transient phenomena in AC motors were clarified and means of eliminating or suppressing them were found to produce very fast responses in motor torque control. Therefore, a numerical solution of the circuit equations is not necessary.

However, for the cases when it may be required, several examples of the numerical solution of performance equations for AC motors, which are written in terms of spiral vectors, will be shown, to indicate that it is easier and less laborious in computer simulation, when the computer program can handle complex numbers.

Equation (4.6) is the circuit equation for current-input control of induction motor torque. It is rewritten as

$$-\tfrac{3}{2}Mpi_1 = R_2 i_2 + (l_2 + \tfrac{3}{2}M)pi_2, \tag{IV.1}$$

whence we get

$$i_2 = -\frac{\tfrac{3}{2}M}{l_2 + \tfrac{3}{2}M} i_1 - \frac{R_2}{l_2 + \tfrac{3}{2}M} \frac{i_2}{p}. \tag{IV.2}$$

Computer simulation circuits for this equation are shown in Figs. IV.1 and IV.2.

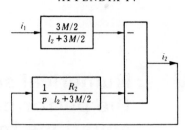

Fig. IV.1   Block diagram of computer simulation of transient response for current-input control of induction motor torque

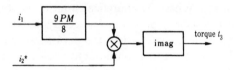

Fig. IV.2   Block diagram of computer simulation of transient torque of the induction motor

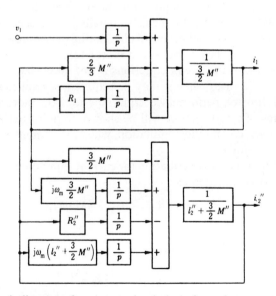

Fig. IV.3   Block diagram of computer simulation of transient response for voltage-input control of induction motor torque

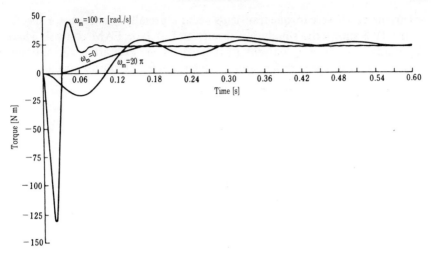

Fig. IV.4 Simulation results of transient torque of the induction motor in Table 3.1 without pre-excitation

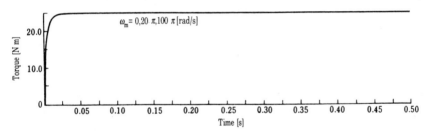

Fig. IV.5 Simulation of results of transient torque of the induction motor in Table 3.1 for T-II type FAM control (with pre-excitation)

For the voltage-input control of induction motor torque, a T-II type equivalent circuit is convenient and its circuit equation is given by equation (5.8), which is modified as

$$i_1 = \frac{1}{\frac{3}{2}M''}\left(\frac{v_1}{p} - R_1\frac{i_1}{p} - \frac{3}{2}M''i_2''\right), \tag{IV.3}$$

$$i_2'' = \frac{1}{l_2'' + \frac{3}{2}M''}\left(-\frac{3}{2}Mi_1 + \frac{3}{2}j\omega_m M''\frac{i_1}{p} - R_2''\frac{i_2''}{p} + j\omega_m(l_2'' + \frac{3}{2}M'')\frac{i_2''}{p}\right). \tag{IV.4}$$

Figure IV.3 shows the computer simulation circuit for these equations. Figure IV.4 shows the simulation results of the induction motor in Table 3.1, for the initial condition of zero currents. Various motor speeds are given as parameters. Severe torque transients occur especially at high motor speeds. Figure IV.5 shows the simulation results of T-II type FAM control, where the steady-state exciting current I_0'' is given as pre-excitation current. The torque responses are very fast and are the same for all motor speeds.

Appendix V   Spiral vector theory of salient-pole synchronous machines

It is widely thought that salient-pole machines can only be analysed by the two-axis theory or d–q axis theory. In the cross section of the synchronous machine of Fig. V-1, the axis along the centre line of the salient magnetic pole on the rotor is called the direct axis and the axis through the rotor centre and perpendicular to it is called the quadrature axis. Along these two axes, the air gap lengths are different and two different inductances ensue. It is often thought that the two-axis theory monopolizes analyses of the salient-pole machine, and its coordinate axes are often called the d and q axes, which are used even for the induction motor with a uniform air gap. However, it seems that the monopoly is superfluous. Now, the salient-pole synchronous machine will be analysed by means of the spiral vector method, in order to show that the method is as good as, or even much better than, the two-axis method.

V.1   Steady-state analysis of the salient-pole synchronous motor

Figure V.1 shows the analytical model of the salient-pole synchronous motor, depicting the phases a and b windings of the three-phase armature winding and the salient-pole rotor. For symmetrical steady-state operation, the variables are expressed as follows. The internal induced voltage of phase a is

$$e_a = \sqrt{2}|\dot{E}_1| \cos(\omega t + \phi) = \mathrm{Re}\left[j\omega\lambda\, e^{j(\omega t + \phi)}\right] = \mathrm{Re}\left[\dot{E}_1\right], \qquad (\text{V.1})$$

where $\dot{E}_1$ is its circular vector, ϕ is its phase angle, and λ is the flux linkage from the magnetic poles. The three-phase currents are

$$i_a = \sqrt{2}|\dot{I}_1| \cos(\omega t + \phi_1) = \mathrm{Re}\left[\sqrt{2}\dot{i}_1\right], \qquad (\text{V.2})$$

$$i_b = \sqrt{2}|\dot{I}_1| \cos(\omega t + \phi_1 - \tfrac{2}{3}\pi) = \mathrm{Re}\left[\sqrt{2}\dot{i}_1\, e^{-\frac{2}{3}j\pi}\right], \qquad (\text{V.3})$$

$$i_c = \sqrt{2}|\dot{I}_1| \cos(\omega t + \phi_1 + \tfrac{2}{3}\pi) = \mathrm{Re}\left[\sqrt{2}\dot{i}_1\, e^{\frac{2}{3}j\pi}\right]. \qquad (\text{V.4})$$

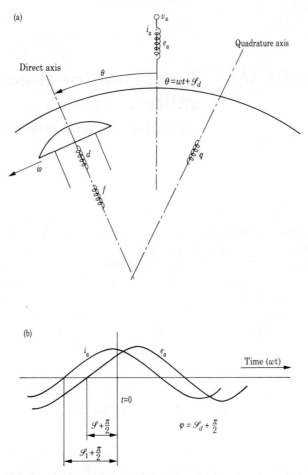

Fig. V.1 Analytical model of the salient-pole synchronous motor, (a) Analytical model (b) Voltage and current waves

As the rotor rotates, the inductances of the armature winding change, in accordance with the following equations. The self-inductances of phases a, b, and c are

$$L_a = L + L' \cos 2\theta = L + L' \cos(2\omega t + 2\phi_d), \tag{V.5}$$

$$L_b = L + L' \cos(2\theta - \tfrac{2}{3}\pi) = L + L' \cos(2\omega t + 2\phi_d - \tfrac{2}{3}\pi), \tag{V.6}$$

$$L_c = L + L' \cos(2\theta + \tfrac{2}{3}\pi) = L + L' \cos(2\omega t + 2\phi_d + \tfrac{2}{3}\pi). \tag{V.7}$$

Mutual inductances between the three phases are

$$M_{ab} = M_{av} + M' \cos(2\theta + \psi) = M_{av} + M' \cos(2\omega t + 2\phi_d + \psi),$$ (V.8)

$$M_{bc} = M_{av} + M' \cos(2\theta + \psi - \tfrac{4}{3}\pi) = M_{av} + M' \cos(2\omega t + 2\phi_d + \psi - \tfrac{4}{3}\pi),$$ (V.9)

$$M_{ca} = M_{av} + M' \cos(2\theta + \psi + \tfrac{4}{3}\pi) = M_{av} + M' \cos(2\omega t + 2\phi_d + \psi + \tfrac{4}{3}\pi).$$ (V.10)

Self- and mutual inductances change as double frequency functions of the power supply frequency ω, which is also the angular speed of the rotor at the synchronous speed in electrical radians/second. The phase angle ψ is the phase difference between variations of L_a and M_{ab}, and it will be determined shortly.

The flux linkage λ_{ga} of phase a due to the rotating field in the air gap is given by

$$\lambda_{ga} = L_a i_a + M_{ab} i_b + M_{ca} i_c + \lambda \cos \theta.$$ (V.11)

The last term is the flux linkage coming from the magnetic pole on the rotor, which is now assumed to be permanent-magnet excited. Inserting equations (V.2)–(V.10) in equation (V.11), we get

$$
\begin{aligned}
\lambda_{ga} = \sqrt{2} |\dot{I}_1| \{ & L \cos(\omega t + \phi_1) \\
& + \tfrac{1}{2} L' \{ [\cos(\omega t + 2\phi_d - \phi_1) + \cos(3\omega t + 2\phi_d + \phi_1)] \\
& + M_{av} [\cos(\omega t + \phi_1 - \tfrac{2}{3}\pi) + \cos(\omega t + \phi_1 + \tfrac{2}{3}\pi)] \\
& + \tfrac{1}{2} M' [\cos(\omega t + 2\phi_d + \psi - \phi_1 + \tfrac{2}{3}\pi) \\
& \qquad + \cos(3\omega t + 2\phi_d + \psi + \phi_1 - \tfrac{2}{3}\pi)] \\
& + \tfrac{1}{2} M' [\cos(\omega t + 2\phi_d + \psi - \phi_1 + \tfrac{2}{3}\pi) \\
& \qquad + \cos(3\omega t + 2\phi_d + \psi + \phi_1)] \} + \lambda \cos \theta
\end{aligned}
$$ (V.12)

In this equation, the triple-frequency terms are

$$
\sqrt{2} |\dot{I}_2| [\tfrac{1}{2} L' \cos(3\omega t + 2\phi_d + \phi_1) + \tfrac{1}{2} M' \cos(3\omega t + 2\phi_d + \psi + \phi_1 - \tfrac{2}{3}\pi) \\
+ \tfrac{1}{2} M' \cos(3\omega t + 2\phi_d + \psi + \phi_1)]
$$ (V.13)

In order for the triple-frequency terms to disappear, the following conditions are sufficient:

$$L' = M', \qquad \psi = -\tfrac{2}{3}\pi.$$ (V.14)

In actual salient-pole machines, these conditions are satisfied.

Under these conditions the flux linkage λ_{ga} of equation (V.12) contains only power frequency terms, as follows:

$$\lambda_{ga} = \sqrt{2}|\dot{I}_1|\{L\cos(\omega t + \phi_1) + M_{av}[\cos(\omega t + \phi_1 - \tfrac{2}{3}\pi) + \cos(\omega t + \phi_1 + \tfrac{2}{3}\pi]$$
$$+ \tfrac{1}{2}L'[\cos(\omega t + 2\phi_d - \phi_1) + \cos(\omega t + 2\phi_d - \phi_1)$$
$$+ \cos(\omega t + 2\phi - \phi_1)]\} + \lambda\cos\theta$$
$$= \sqrt{2}|\dot{I}_1|\{(L - M_{av})\cos(\omega t + \phi_1) + \tfrac{3}{2}L'\cos(\omega t + 2\phi_d - \phi_1)\} + \lambda\cos\theta.$$

$$(V.15)$$

Since $M_{av} = L\cos\tfrac{2}{3}\pi = -\tfrac{1}{2}L$, this equation becomes

$$\lambda_{ga} = \sqrt{2}|\dot{I}_1|\{\tfrac{3}{2}L\cos(\omega t + \phi_1) + \tfrac{3}{2}L'\cos(\omega t + \phi_1 + 2\phi_d - 2\phi_1)\} + \lambda\cos(\omega t + \phi).$$

$$(V.16)$$

In terms of circular vectors, this equation becomes

$$\lambda_{ga} = \sqrt{2}|\dot{I}_1|\{\tfrac{3}{2}L\ e^{j(\omega t + \phi_1)} + \tfrac{3}{2}L'\ e^{j(\omega t + \phi_1)}\ e^{j(2\phi_d - 2\phi_1)}\} + \lambda\ e^{j(\omega t + \phi_d)}. \quad (V.17)$$

Hence,

$$\lambda_{ga} = \tfrac{3}{2}Li_a + \tfrac{3}{2}L'i_a\ e^{j(2\phi_d - 2\phi_1)} + \lambda\ e^{j(\omega t + \phi_d)}, \quad (V.18)$$

where i_a is the circular vector

$$i_a = \sqrt{2}|\dot{I}_1|\ e^{j(\omega t + \phi_1)}. \quad (V.19)$$

The circuit equation of phase a is now given by

$$v_a = R_1 i_a + l_1 p i_a + p\lambda_{ga} = R_1 i_a + l_1 p i_a + (\tfrac{3}{2}L + \tfrac{3}{2}L'\ e^{j(2\phi_d - 2\phi_1)})p i_a + e_a. \quad (V.20)$$

Here, the internal induced voltage e_a is

$$e_a = j\omega\lambda\ e^{j(\omega t + \phi_d)} = \sqrt{2}\dot{E}_1\ e^{j(\omega t + \phi)}, \qquad \phi = \phi_d + \frac{\pi}{2}. \quad (V.21)$$

The circuit equation (V.20) contains only the state variables of phase a, which is segregated from the other phases. Subscript a is now changed to 1. Equation (V.20) then becomes

$$v_1 = R_1 i_1 + l_1 p i_1 + (\tfrac{3}{2}L - \tfrac{3}{2}L'\ e^{j(2\phi - 2\phi_1)})p i_1 + e_1. \quad (V.22)$$

The inductance for this equation is

$$l_1 + \tfrac{3}{2}L - \tfrac{3}{2}L'[\cos(2\phi - 2\phi_1) + j\sin(2\phi - 2\phi_1)], \quad (V.23)$$

which is a complex inductance. When $\phi - \phi_1 = \pm\pi/2$, it takes the following largest real value:

$$l_1 + \tfrac{3}{2}L + \tfrac{3}{2}L' = L_d, \quad (V.24)$$

which is called the direct-axis inductance. When $\phi - \phi_1 = 0$, it takes the following smallest real value:

$$l_1 + \tfrac{3}{2}L - \tfrac{3}{2}L' = L_q, \quad (V.25)$$

which is called the quadrature-axis inductance.

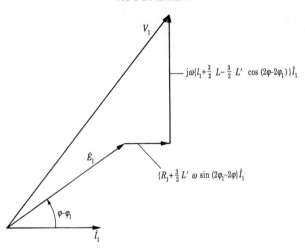

Fig. V.2   Circular vector diagram at $t = 0$ of the salient-pole synchronous motor

For the steady state, the state variables become circular vectors and equation (V.22) becomes

$$\dot{V}_1 = R_1\dot{I}_1 + j\omega[l_1 + \tfrac{3}{2}L - \tfrac{3}{2}L'\cos(2\phi - 2\phi_1) - \tfrac{3}{2}jL'\sin(2\phi - 2\phi_1)]\dot{I}_1 + \dot{E}_1$$
$$= R_1\dot{I}_1 + \tfrac{3}{2}\omega L'\sin(2\phi - 2\phi_1)\dot{I}_1$$
$$+ j\omega[l_1 + \tfrac{3}{2}L - \tfrac{3}{2}L'\cos(2\phi - 2\phi_1)]\dot{I}_1 + \dot{E}_1. \quad (V.26)$$

The corresponding vector diagram for $t = 0$ is shown in Fig. V.2. The resistance is increased by $\tfrac{3}{2}\omega M'\sin(2\phi - 2\phi_1)$ because of saliency, and the motor torque due to the saliency is

$$\tfrac{9}{4}L'P\sin(2\phi - 2\phi_1)|\dot{I}_1|^2 \qquad [\text{N m}], \qquad (V.27)$$

which is called saliency torque or reluctance torque.

V.2   Transient-state analysis of the salient-pole synchronous motor

For the transient state, the state variables are represented by spiral vectors. The three-phase armature currents under symmetrical operation are then expressed as follows:

$$i_a = A\,e^{\delta t} = A\,e^{-\lambda t}\,e^{j\omega't}, \qquad i_b = A\,e^{\delta t - \frac{2}{3}j\pi}, \qquad i_c = A\,e^{\delta t + \frac{2}{3}j\pi}. \quad (V.28)$$

Here, $\delta = -\lambda + j\omega'$ is a characteristic root. Their instantaneous real values are as follows:

$$
\left.\begin{aligned}
i_a &= |A|\, e^{-\lambda t} \cos(\omega' t + \phi_1),\\
i_b &= |A|\, e^{-\lambda t} \cos(\omega' t + \phi_1 - \tfrac{2}{3}\pi),\\
i_c &= |A|\, e^{-\lambda t} \cos(\omega' t + \phi_1 + \tfrac{2}{3}\pi).
\end{aligned}\right\} . \tag{V.29}
$$

The gap flux linkage of phase a is given by equation (V.11), under the conditions of equation (V.14), as follows:

$$
\begin{aligned}
\lambda_{ga} &= \sqrt{2}|A|\, e^{-\lambda t}[L_a \cos(\omega' t + \phi_1) + M_{ab}\cos(\omega' t + \phi_1 - \tfrac{2}{3}\pi)\\
&\qquad + M_{ca}\cos(\omega' t + \phi_1 + \tfrac{2}{3}\pi)] + \lambda \cos(\omega t + \phi_d)\\
&= \tfrac{3}{2}L|A|\, e^{-\lambda t}\cos(\omega' t + \phi_1) + \tfrac{3}{2}L'|A|\, e^{-\lambda t}\cos[(2\omega - \omega')t + 2\phi_d - \phi_1]\\
&\qquad\qquad\qquad + \lambda \cos(\omega t + \phi_d). \tag{V.30}
\end{aligned}
$$

Expressed in terms of spiral vectors, this equation becomes

$$
\begin{aligned}
\lambda_{ga} &= \tfrac{3}{2}L\, e^{-\lambda t}|A|\, e^{j(\omega' t + \phi_1)} + \tfrac{3}{2}L'\, e^{-\lambda t}|A|\, e^{j[(2\omega - \omega')t + 2\phi_d - \phi_1]} + \lambda\, e^{j(\omega t + \phi_d)}\\
&= \tfrac{3}{2}Li_a + \tfrac{3}{2}L'\, e^{j(2\omega t + 2\phi)} i_a^* + \lambda\, e^{j(\omega t + \phi_d)}. \tag{V.31}
\end{aligned}
$$

Here, i_a^* is the complex conjugate of i_a. The circuit equation of phase a is now given by

$$
\begin{aligned}
v_a &= R_1 i_a + l_1 p i_a + p\lambda_{ga} + e_1\\
&= R_1 i_a + (l_1 + \tfrac{3}{2}L)p i_a + \tfrac{3}{2}L' p(e^{j(2\omega t + 2\phi_d)} i_a^*) + e_1. \tag{V.32}
\end{aligned}
$$

This equation contains variables of phase a only, and in order for phase a to represent the primary circuit, subscript a is changed to 1. Then equation (V.32) becomes

$$
v_1 - e_1 = R_1 i_1 + (l_1 + \tfrac{3}{2}L)p i_1 + (\tfrac{3}{2}L')p(e^{j(2\omega t + 2\phi_d)} i_1^*). \tag{V.33}
$$

The general solution of this equation is

$$
i_1 = i_{1t} + \sqrt{2}|\dot{I}_1|\, e^{j(\omega t + \phi_1)}. \tag{V.34}
$$

The second term is the steady-state term obtained in the previous section. Inserting i_1 of equation (V.34) in equation (V.33), we get

$$
0 = R_1 i_1 t + (l_1 + \tfrac{3}{2}L)p i_{1t} + (\tfrac{3}{2}L')p(e^{j(2\omega t + 2\phi_d)} i_{1t}^*). \tag{V.35}
$$

The last term is explicitly a time function, which acts as a forcing function. Let us assume i_{1t} to be

$$
i_{1t} = A\, e^{-\lambda t} \tag{V.36}
$$

for the forcing function. Then equation (V.35) becomes

$$R_1 i_{1t} + (l_1 + \tfrac{3}{2}L)p i_{1t} = -\tfrac{3}{2}L'(-\lambda + 2j\omega)A\, e^{(-\lambda + 2j\omega)t + 2j\phi_d}. \quad \text{(V.37)}$$

The general resolution of this equation is given by the formula of equation (1.16) as

$$i_{1t} = A\, e^{-\lambda t} - \frac{\tfrac{3}{2}L'(-\lambda + 2j\omega)}{R_1 + (-\lambda + 2j\omega)(l_1 + \tfrac{3}{2}L)}\, A\, e^{(-\lambda + 2j\omega)t + 2j\phi_d}, \quad \text{(V.38)}$$

where $\lambda = R_1/(l_1 + \tfrac{3}{2}L)$. The first term is the assumed solution of equation (V.36) and the second additional term is a double-frequency term. Since the second term also contains A and is much smaller in magnitude than the first term, i_{1t} of equation (V.38) can be considered to be a correct transient state solution. If necessary, it is possible to derive a more accurate solution by inserting i_{1t} of equation (V.38) into the right-hand side of equation (V.37) and repeating the same process as above. Then the characteristic root becomes complex, replacing real λ in equation (V.38). However, its imaginary part is so small that λ in equation (V.38) is good enough in most cases.

Inserting i_{1t} of equation (V.38), we obtain the general solution of equation (V.34) as

$$i_1 = A\, e^{-\lambda t} - \frac{\tfrac{3}{2}L'(-\lambda + 2j\omega)}{R_1 + (-\lambda + 2j\omega)(l_1 + \tfrac{3}{2}L)}\, A\, e^{(-\lambda + 2j\omega)t + 2j\phi_d} + \sqrt{2}|\dot{I}_1|\, e^{j(\omega t + \phi_1)}.$$
$$\text{(V.39)}$$

All terms are spiral vectors: the first term is a decaying DC, the second term is a decaying double-frequency term, and the last term is a steady-state term. The general solution contains an arbitrary constant A, which is to be determined by an initial condition. If, at $t = 0$, the condition

$$i = \sqrt{2}|\dot{I}_1|\, e^{j\phi_1} \quad \text{(V.40)}$$

is satisfied, then A becomes zero and no transient occurs. Equation (V.40) is the transientless condition for the current input control of the synchronous motor.

When the current is not kept constant in the field winding and a transient current flows, the analysis becomes a little more complicated. But a similar approach using the spiral vector method is also applicable to this case.

V.3 Computer simulation of the synchronous machine

Equation (V.33) is the performance equation of a synchronous machine seen as a motor, and can be modified as

$$i_1 = \frac{v_1 - e_1}{pL_s} - \frac{R_1}{pL_s} i_1 - \frac{3L'}{2L_s}(e^{2\theta}i_1^*)$$

$$L_s = l_1 + \tfrac{3}{2}L. \quad \text{(V.41)}$$

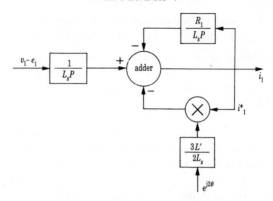

Fig. V.3 Computer simulation of the salient-pole synchronous machine under symmetrical operation

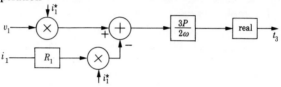

Fig. V.4 Computer simulation of synchronous machine torque for equation (V.42)

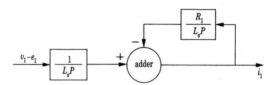

Fig. V.5 Computer simulation of non-salient-pole synchronous machines

Figure V.3 shows the block diagram for a computer simulation of equation (V.41), and Fig. V.4 shows that for motor torque, given by

$$t_3 = \frac{3P}{2\omega} \text{real}[v_1 - Ri_1]i_1^* \tag{V.42}$$

For nonsalient-pole machines, L' becomes zero in equation (4.40), and Fig. V.3 becomes Fig. V.5, which does not contain a timely variant coefficient, and is very simple.

V.4   Transient analysis of the salient-pole synchronous machine under asymmetrical operation

As explained in Section 10.4, the spiral vector method extends the applicability of the symmetrical component method to the transient analysis of asymmetrical (or unbalanced) operation of three-phase machines. In the asymmetrical operation of a three-phase synchronous machine there are three characteristic roots, which are assigned to three independent positive-, negative-, and zero-sequence components. Under symmetrical operation, the armature currents are given by equation (V.29), which shows their positive-sequence components. Under asymmetrical operation, armature currents contain negative-sequence components which are expressed as

$$
\left.\begin{aligned}
i_{a2} &= \sqrt{2}|\dot{I}_2|\, e^{-\lambda_2 t} \cos(\omega' t + \phi_2) \\
i_{b2} &= \sqrt{2}|\dot{I}_2|\, e^{-\lambda_2 t} \cos(\omega' t + \phi_2 + \tfrac{2}{3}\pi) \\
i_{c2} &= \sqrt{2}|\dot{I}_2|\, e^{-\lambda_2 t} \cos(\omega' t + \phi_2 - \tfrac{2}{3}\pi).
\end{aligned}\right\}
\tag{V.43}
$$

Inserting these values, equation (V.11) becomes

$$
\lambda_{ga2} = (\tfrac{3}{2}L)\, e^{-\lambda_2 t}\sqrt{2}|\dot{I}_2| \cos(\omega' t + \phi_2) + (\tfrac{3}{2}L')\, e^{-\lambda_2 t} \cos(\overline{2\omega + \omega' t + 2\phi + \phi_2}).
\tag{V.44}
$$

where λ_{ga2} is the negative-sequence component of the magnetic flux linkage of phase a. Using spiral vectors, λ_{ga2} becomes

$$
\lambda_{ga2} = (\tfrac{3}{2}L)i_{a2} + (\tfrac{3}{2}L')\, e^{j2\theta}i_{a2}.
\tag{V.45}
$$

The circuit equation of phase a is now given by

$$
\begin{aligned}
v_{a2} &= R_1 i_{a2} + l_1 p i_{a2} + p\lambda_{ga2} \\
&= R_1 i_{a2} + (l_1 + \tfrac{3}{2}L)p i_{a2} + (\tfrac{3}{2}L')p(e^{2j\theta}i_{a2}).
\end{aligned}
\tag{V.46}
$$

This equation contains state variables only of phase a, which is segregated from the other phases. The subscript a is now omitted, giving

$$
v_2 = R_1 i_2 + L_s p i_2 + (\tfrac{3}{2}L')p(e^{j2\theta}i_2).
\tag{V.47}
$$

This equation now contains only negative-sequence components and is thus normalized in the mathematical sense. Comparing this with equation (V.33) for positive-sequence components, the following differences are noticed. There is no internal induced voltage e_2 and in the last term, i_1^* is replaced by i_2. A general solution can be similarly derived, but we omit its derivation here. It should be pointed out that equation (V.47) generates triple frequency currents under steady-state conditions.

Equation (V.47) is modified as

$$
i_2 = \frac{v_2}{pL_s} - \frac{R_1}{pL_s}i_2 - \frac{3L'}{2L_s}(e^{j2\theta}i_2).
\tag{V.48}
$$

Figure V.6 shows the block diagram of computer simulation of equation (V.48).

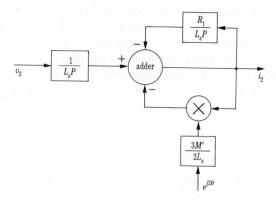

Fig. V.6 Computer simulation of the salient-pole synchronous machine for negative-sequence components

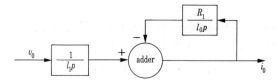

Fig. V.7 Computer simulation of synchronous machines for zero-sequence components

It is well known that zero-sequence component currents do not produce gap flux, irrespective of whether the machine is of the salient- or nonsalient-pole type, and the circuit equation for these is given by

$$v_0 = R_1 i_0 + l_1 p i_0. \qquad (V.49)$$

Its computer simulation is shown in Fig. V.7.

The following remarks should be made. For nonsalient synchronous machines, L' becomes zero in all equations and computer simulations, making them much simpler and producing analytical results that are in agreement with the analyses in Chapter 7.

Although the machines are here treated as synchronous motors, these equations are also valid for generators if the signs of the state variables are appropriately changed.

The spiral vector method explained here considerably facilitates transient and stability analyses of power systems incorporating synchronous machines.

Appendix VI Derivation of transient-state solutions for synchronous motor torque

Transient-state currents in a synchronous motor with a damper winding were obtained as equations (7.72)–(7.75). Now the transient-state torque will be obtained by inserting these equations into equations (7.54) and (7.55). The characteristic roots δ_1 and δ_2 are modified as follows:

$$\delta_1 = -\lambda_1 + j\omega_1, \qquad \delta_2 = -\lambda_2 + j\omega_2. \tag{VI.1}$$

The following relation can be obtained from equation (7.70):

$$\omega_1 + \omega_2 = \omega. \tag{VI.2}$$

This is the same as equation (4.47) when $\omega = \omega_m$.

The following relations are introduced:

$$\delta_1 - \delta_2 = \lambda_2 - \lambda_1 + j(\omega_1 - \omega_2) = [(\lambda_2 - \lambda_1)^2 + (\omega_1 - \omega_2)]^{\frac{1}{2}} e^{j\xi_0}, \tag{VI.3}$$

$$\delta_1 - j\omega = -\lambda_1 - j\omega_2 = -(\lambda_1^2 + \omega_2^2)^{\frac{1}{2}} e^{j\xi_1}, \tag{VI.4}$$

$$\delta_2 - j\omega = -\lambda_2 - j\omega_1 = -(\lambda_2^2 + \omega_1^2)^{\frac{1}{2}} e^{j\xi_2}, \tag{VI.5}$$

$$R_2 + \tfrac{3}{2}M(\delta_1 - j\omega) = -\tfrac{3}{2}M\lambda_1 - R_2 + \tfrac{3}{2}Mj\omega_2)$$
$$= -[(\tfrac{3}{2}M\lambda_1 - R_2)^2 + (\tfrac{3}{2}M\omega_2)^2]^{\frac{1}{2}} e^{j\xi_3}, \tag{VI.6}$$

$$R_2 + \tfrac{3}{2}M(\delta_2 - j\omega) = -(\tfrac{3}{2}M\lambda_2 - R_2 + \tfrac{3}{2}Mj\omega_1)$$
$$= -[(\tfrac{3}{2}M\lambda_2 - R_2)^2 + (\tfrac{3}{2}M\omega_1)^2]^{\frac{1}{2}} e^{j\xi_4}, \tag{VI.7}$$

$$\left.\begin{array}{ll} \gamma_1 = \xi_2 + \xi_3 - \xi_0, & \gamma_2 = \xi_1 + \xi_4 - \xi_0, \\ \gamma_3 = \xi_1 + \xi_2 - \xi_0, & \gamma_4 = \xi_2 - \xi_0, \qquad \gamma_5 = \xi_1 - \xi_0. \end{array}\right\} \tag{VI.8}$$

With these equations, equations (7.50)–(7.55) give the transient-state torque solutions of equations (7.76) and (7.77).[7]

References

1. Yamamura, S. and Nakagawa, S. (1980). Transient analysis and control of AC servomotor—Proposal of field acceleration method. *Trans. IEE Japan*, **B-101**, 557.
2. Yamamura, S. and Nakagawa, S. (1981). Equivalent circuit and field acceleration method control of AC servomotor of induction motor type. *Trans. IEE Japan*, **B-102**, 439.
3. Yamamura, S., Ka, S. and Nakagawa, S. (1982). Equivalent circuit of induction motor as servomotor of quick response. *Trans. IEE Japan*, **B-103**, 133.
4. Yamamura, S., Ka, S., and Kawamura, A. (1983). Equivalent circuit of induction motor as servomotor of quick response. *Proc. IPEC IEE of Japan*, 732.
5. Yamamura, S., Ka, S., Nakagawa, S., and Kawamura, A. (1982). Transient analysis and field acceleration method control of induction motor. *Trans. IEE Japan*, **B-103**, 491.
6. Yamamura, S. and Nakagawa, S. (1983). Voltage input FAM control of induction motor. *Trans. IEE Japan*, **B-104**, 449.
7. Yamamura, S. and Kondo, K. (1983). Quick response of synchronous motor as control motor. *Trans. IEE Japan*, **B-104**, 763.
8. Yamamura, S. (1985) Transient analysis of AC machine by means of spiral vector method. *Trans. IEE Japan*, **B-105**, 581.
9. Yamamura, S. (1986). Modern theory of ac motors: Phase segregation method, d.r. vector method and field acceleration method. *Proc. INCEMDAS '86*.
10. Yamamura, S. (1986). *AC Motors for High Performance Applications*. Dekker.
11. Yamamura, S. (1987). Modern theory of AC motors—Analysis and control. *Proc. BICEM, IEE of China*.
12. Yamamura, S., Nakamori, K., and Takano, S. (1987). Secondary current feedback control of induction motor torque. *Proc. Ann. Conf., IEE of Japan*, 783.
13. Takano, S., Nakamori, K., and Yamamura, S. (1986). Computer simulation in terms of spiral vectors of voltage-input control of induction motor torque. *Proc. Ann. Conf., IEE of Japan*, No. 681.
14. *Fundamental Electrical Machinery Theory* (1984). IEE of Japan.
15. *Electrical Machinery* (1985). IEE of Japan.
16. Bekku, T. (1927). On short circuit current of ac generator. *J. IEE Japan*, p. 805.
17. Fortescue, C. L. (1918). Method of symmetrical co-ordinates applied to the solution of polyphase networks. *Proc. 34th Ann. Conv. of AIEE*. 1027.
18. Lyon, W. V. (1954). *Transient Analysis of Alternating Current Machinery*. Wiley.
19. Kovacs, P. K. (1984). *Transient Phenomena in Electrical Machines*, Elsevier.

20. Yamamura, S. (1989). *Analysis and Control of AC Motors*, Ohmsha, Tokyo.
21. Yamamura, S. (1988). Theory of ac motor analysis and control *Proc. ICEM*, **1**, 1.
22. Yamamura, S. (1988). Secondary current feedback control of induction motor. *Proc. ICEM*.
23. Yamamura, S. (1989). Spiral vector theory of ac circuit and ac machine. *Proc. Japan Academy* 65, Ser. B, No. 6, 142.
24. Yamamura, S. (1990). Spiral vector theory of ac motor analysis and control. *Proc. APEC '90 of IEEE*, 77.
25. Vas, P. (1990). *Vector control of AC machines*, Oxford University Press.
26. Yamamura, S. (1991). Spiral vector method and symmetrical component method. *Proc. Japan Academy*, 67, Ser. B, No. 1, 1–6.
27. Yamamura, S. (1992). Spiral vector theory of salient-pole synchronous machines. *Proc. Japan Academy*, 67, Ser. B, No. 3.

Index